DLSF

By

Willis J. Pierre

With special thanks to
All My Mathematics Students

Library of Congress Card Number: 2003097495

ISBN 0-9742936-0-1

First printing   October 2003

Printed in the U.S.A. by
Morris Publishing
3212 East Highway 30
Kearney, NE 68847
1-800-650-7888

Dedicated to: All DLSFs

wherever they may be

Table of Contents

Contents

Introduction

For forty-one years I taught mathematics in private secondary schools. Although I have received many grants and awards the only criticism I have received is that I have done little to help other teachers. Any help I have given is either by example or by answering specific questions the teacher initiates. Therefore, I wrote this book for the purpose of helping others. First of all, I hope *DLSF* may inspire some readers to become teachers. Secondly, I hope to inspire some teachers to enter the field of mathematics. Finally, I hope the remaining readers will have a greater understanding of the teaching profession.

Some, I am sure, are going to accuse me of boasting. I apologize in advance if I give that appearance. I am proud of what I have achieved. Any honors or awards I have received I detail here as part of the story. I have accomplished nothing more than any young person starting out today can accomplish. I hope a few young college students will be inspired by what they read in this book, to enter the teaching profession and accomplish infinitely more than I could have ever dreamed of doing.

Nobody can spend forty-one years in the teaching profession without having many, many people to thank. I cannot thank them all by name as there would not be any pages left for the book and what I want to say. However, I have been aided in so many ways by colleagues on the faculties and the administrators in my schools. You, the reader, will get to meet many of these people as you travel through *DLSF*.

I also must thank the students I have had in class. They have been an inspiration for me to teach and to teach well. So often I have been in awe of what they have said and done. My great gratitude also goes to the parents of my students who so graciously shared with me their most prized possessions. Having taught in private boarding schools I have been *in loco parentis* for their sons and daughters. I hope that I have carried out those responsibilities satisfactorily.

To teach is a real joy; to share the accomplishments and failures of so many people. A teacher and his students bond in a unique way. I am blessed to be able to call so many of my former students friends.

Since this is a math book, answers to problems are provided in Appendix A.

I hope every reader enjoys *DLSF* as much as I have enjoyed writing it.

Willis J. Pierre

Chapter I

Lake Forest Academy Years

The plane seemed to be flying quite low. Looking out the window, I could easily see everything on the ground. Although the cars were very small, I could decipher the Pennsylvania Turnpike winding its way westward through the mountains of the western part of that state. It was an astonishingly clear day, not a cloud in the sky. The bright sun lit up and brought into focus everything on the ground.

This was my first time in an airplane. That seems strange by today's standards that at age twenty-two I was taking my first airplane flight. I had an interview for a teaching job at Lake Forest Academy in Illinois. That school had offered to pay my way to come to its campus for a visit and interview. What did I have to lose? Who knows, it might be a more interesting and exciting job than the offers I already had on my desk. The plane left Philadelphia International Airport early in the morning headed to Chicago's O'Hare Airport. I had never been to Chicago and was excited to see the city as well.

There seemed to be nothing really unusual as we took off although I had a weird feeling in my stomach and seemed to be quite dizzy as we glided west. I was happy I had a window seat and a great view because the day was so clear. I attributed my illness to nerves. I was nervous because of the interview and also

1

about where and how I would meet the headmaster of Lake Forest Academy. I had not noticed the plane changing direction, but I suddenly became aware that we were flying eastward as the sun was in front of us. Just as I recognized this situation, the captain's voice spoke on the intercom. "We are having some difficulty and are returning to Philadelphia." That was all that was said. I surmised that if it were something really threatening he would have said more. As we approached the runway in Philadelphia, I could see many fire engines and ambulances lining the runway. Luckily, I was new to the flying experience and thought this was a normal precaution. We landed without incident and never learned what the problems were.

After a two-hour wait we were able to take off again. This time the trip was quite successful. I had a sandwich and cup of coffee and kept thinking about that interview. Next to me was a man from Evanston, Illinois, who with his friendly conversation quieted my nerves. As we got off the plane he handed me a piece of paper with his name and phone number. He asked that if I landed the job, I call him when I arrived at school in September. I hate to admit that I never followed up on that offer. I wish I had. I did keep that piece of paper for several years but I never had the courage to call the number. I am sure he thought I never got the job and never gave it another thought. However, that phone call could have opened some doors. My first piece of advice to the reader is never to close a door or leave one closed that is easily opened. You never know where it will lead. You can always turn around and go back out if you do not like what you see inside.

At the airport the first thing I did was call Lake Forest Academy. The headmaster had received word that the plane would be at least two hours late. He told me to stand outside the terminal where the plane had landed and look for a brown station wagon with a Lake Forest Academy pennant waving from the radio antenna. He promised that he would be there in one-half hour. The headmaster arrived exactly one-half hour later. I

2

hopped in the front seat and introduced myself. I was so young and naïve that I had not even brought a briefcase with me, but he had my credentials which the teacher placement office had sent.

After getting into his car we made small talk as he too could see I was nervous, having just endured my first airplane ride. Arriving at the School I was impressed with the entrance to the campus. There was an entrance gate and then a long winding road through trees and past a large lake. It was beautiful. We arrived at the headmaster's home where I was introduced to his wife, Florie. She was just as impressive as her husband. Harold Corbin was a gifted man who had been headmaster of Lake Forest Academy for several years and had a first-rate plan for making his school into one that was equal in quality to the reputable eastern boarding schools. He was both an academician and an athlete. Lake Forest Academy at that time was an all-boy's school with an enrollment of 225 in grades nine through twelve and he wanted his boys to excel in academics and athletics. He made it clear that at his school teachers were to be involved with the teaching of the entire student. Good habits and moral fiber were to be stressed at all times and proper behavior was always to be expected.

We toured the empty campus. The school year had ended and the entire faculty had left for their summer vacations. I was disappointed that I was not going to get to meet anyone other than the headmaster. However, if the other faculty members were anything like their leader, this school was solid. We ended up in the Headmaster's Office for the formal interview. While interviewing me Harold Corbin admitted he was looking through another applicant's file and remarked this was quite unusual. To this day I do not know what he was checking in that folder. I suspect that he was trying to get my reaction. At one point he made a comment about my receding hairline. I am positive again that he wanted to see my reaction. He knew that students very well may joke about a twenty-two year old who was going bald

3

and he wanted to see how I would react. The one statement that really bothered me however was that he expected every new teacher to stay at least two years. He told me that a new teacher is useless in his first year. That teacher has to learn the ropes and make all the usual first-year teacher mistakes. I thought to myself, but you do not know me. I am not going to make those mistakes. I am going to be an outstanding teacher from day one. (I guess I was very naïve in this area also. I often look back feeling sorry for those students I had in class my first couple years. I made mistakes, but I think I learned from every mistake I made.) I had to respond to Mr. Corbin's two-year requirement. I was not positive I wanted to be a teacher. I thought I did and I knew I wanted to try. However, I was not about to make a two-year commitment. I was not going to give my word I would stay for two years. The fact that your word means something and is like a signed contract would be brought home to me again in later years.

Luckily, Harold Corbin, being a great headmaster and diplomat, understood my feelings. We reached an agreement that, if he offered a position and I accepted it, I would teach at Lake Forest Academy for one year. If I did not like teaching, I could leave at the end of that year, but I could not leave to take a teaching job at another school. Another statement he made has really stuck with me over the years. I think it is unique. He stated that my first two raises would be only token raises. It was up to me to prove that I was a good teacher. After that he would prove to me that he did not want to lose me. Here was a man who in a few minutes convinced me of his honesty and sincerity and so I was convinced that he would follow through on that promise. And he did live up to his word. My first two raises were $100 per year. After that the raise was usually fifteen percent per year. He was truly a man who kept his end of the bargain and I respected him for that.

Later that afternoon we returned to the headmaster's home for dinner. We sat in the parlor and he asked whether I would like a Scotch. I said yes because I did not think it wise to turn his offer down. I was not a drinker and did not really enjoy alcohol. I did manage to consume the one glass. The main course was pork chops which offered a slight problem for me. I am a very fussy eater and pork is one of my least favorite meats. I managed to get a pork chop down also and left after dinner for the airport. A school driver took me back to the airport and we had a good chat. Looking back, I am sure that he was told to ask many of his questions and to relay my answers to the headmaster. I assured him that I was very impressed with Lake Forest Academy and would probably accept, if offered, the position. The following morning I had a phone call from Harold Corbin stating that he had mailed a contract to begin teaching at Lake Forest Academy that September. I accepted verbally on the phone and told him I would sign and return the contract immediately.

Now it was time to think of beginning my teaching career in a summer school. I had accepted a position to teach math for eight weeks at Brick House School, a very small all-girl's school outside Philmont, New York. That summer there were twenty students. Mrs. Burleigh was the owner and headmistress. She had a housemother and two faculty members, of whom I was one. The other faculty member was George Gauthier, a graduate student at the time, working on his doctorate at Princeton University. He was a master of foreign languages, particularly French. I believe he taught four different languages that summer. Mrs. Burleigh also had a cook, a young woman who lived just down the street and came up to prepare meals every day except Sunday. Every Sunday Mrs. Burleigh was the cook. Also on that small campus was a houseboy, Larry, who had just graduated from high school. He lived in town and would arrive early each morning and leave after dinner each evening. He was responsible for mowing the lawn, cleaning the pool, and helping in the kitchen to prepare meals and wash dishes.

The summer was very good for me in many ways. First of all, I made two very close life-long friends, George Gauthier and Larry. As the only males on the campus, we formed a very close bond. George had to teach in the afternoons and was not as free as I was. I taught only morning classes. Larry also was given a couple hours off every afternoon. Often he and I took off to his house or just explored the surrounding wooded area. Larry also invited both George and me to his home many evenings. His parents were most gracious and seemed to appreciate the relationship Larry had with the two of us. To this day, over forty years later, we continue to stay in close touch. Only once after that remarkable summer have all three of us been able to get together. Several years ago I was able to arrange such a reunion. We may have been thirty-five years older than when we first met, but it seemed that nothing had changed. We recalled many stories and memories. Larry later attended the Naval Academy at Annapolis where he was called by his first name instead of his middle name, Larry. So the reader probably knows my friend Larry as Ollie North. I have returned to Philmont frequently trying to relive that summer experience. Of course, without Larry and George, my days in Philmont can never be the same. Another highlight of the summer besides the classroom experience was the proximity of this school to music concerts at Tanglewood and ballet at Jacob's Pillow. There were also several neat trips to Stockbridge, Massachusetts for summer theatre. I learned much from my first teaching job at Brick House that I could apply to my first fall term of secondary school teaching.

I made many of my beginner's mistakes that summer and hoped I would not repeat them when I started my teaching career at Lake Forest Academy in the fall. I learned that summer the difference between being able to solve math problems and explaining to others how to solve them. The important aspect in solving a problem is not the solution itself, but rather the methods and approaches used to find the solution. That summer I always

6

outlined what I would teach the next day in class. This outline included notes for the new material and all sample problems I would use in class. This is a practice I continued throughout my career. Every year a couple students would comment to me that I should save the notes and use them again the following year. I always felt that the next time teaching that material I could do it better. The teacher who uses the same thing over and over again for twenty-five years has not had twenty-five years teaching experience, but rather one year of experience twenty-five times. September came around fast and I was about to embark on what would be a fantastic and fun-filled life of teaching. I drove my old 1952 Plymouth from Princeton, New Jersey, to Lake Forest, Illinois. It was packed with all my earthly possessions. I had not seen my apartment during the interview. It was small, very small. There was a living room, a bedroom, a bath, and a closet for a kitchen. This closet had a refrigerator and hotplate. That was it! At the time I was thrilled. That was not bad for a first-year teacher. At least it was free and I did not have to go out and rent an apartment and I did not have to cook as I could take all my meals in the Academy's dining room. At that time, boarding schools also provided maid service, so I never had to clean my apartment since the cleaning lady came into the apartment every day and straightened things up.

That first day I met Sid Ainsworth, who was Chairman of the Mathematics Department. He was a very tall, thin man about 50 years old. He was a school man in every sense of the word. He loved teaching and he loved his boys, both the students he had in class and those he had in his dormitory. I was very fortunate to have such a strong leader as my department chair in my first year. I brought every test to him for his critique before I administered it. He was a master teacher. His pre-calculus course did not use a text. He taught it from the top of his head and it was up to the students to take good notes. He structured the course in such a way that every student was able to take the notes and end the year with what could easily pass for a textbook. I certainly learned

7

from him to take chances. If you think you can do something that would be a meaningful learning experience for the students, try it. It can work only if you give it a try. Be sure to put your all into it and make sure that you have given it your best effort.

I lived in a dormitory where Sid Ainsworth also lived. This was fortunate in that it gave me easy access to him. He took me under his wing and I owe him a lot for anything I have accomplished as a teacher. On his retirement, he was interviewed for the School's Alumni Bulletin. Although I was no longer at Lake Forest, I did receive copies of its publications. I read that article one night and wanted to cry as I learned all Sid Ainsworth had accomplished in his teaching career. And then came one comment which did bring a tear to my eye. He was asked for some of the great remembrances he had teaching at Lake Forest Academy. He stated that he had the privilege of teaching with Willis Pierre. Several years later that same publication at Lake Forest Academy did a booklet on the great teachers of the School. It included a disclaimer to the effect that before the editors were bombarded by alumni for not including Sid Ainsworth and Willis Pierre, the Bulletin used only those faculty who taught for at least twenty years at the Academy.

During my second year at Lake Forest Academy Sid allowed me to teach the Advanced Placement Calculus course. I spent a great amount of time that summer getting prepared to teach the calculus. At the time, I *thought* I did a great job and the students did reasonably well on the Advanced Placement Tests. However, within a couple years I *knew* I was doing so much better and students were scoring extremely well on the Advanced Placement Tests.

During my first year, I taught a geometry course. I had two students in the class who were constantly coming to me for extra help. With all the extra tutoring they were able to be quite successful in geometry. It has been said that teachers probably learn as much from students as students learn from teachers. In the case of these two geometry students, that statement is in fact true. In the spring they came in to say that I had given them so much of my time that they were going to teach me how to play tennis. I love all sports, but summer was always baseball time. It was almost impossible to get together 18 people to play a game of baseball, but it would be easy to find another person for a game of tennis. These two students worked in the summer at a tennis club as assistants to a professional tennis instructor. They were good teachers. They were able to convince me to enjoy the game and they even taught me to hit a reasonable backhand. I think they spent as many hours teaching me tennis as I spent teaching them geometry. Until a few years ago, when I began to have eye problems, tennis was a game that occupied much of my time. In fact, seldom did a day go by in the summer time that I did not play tennis. There was also a period that I played in a mixed doubles group all winter. It is a great game and also one in which a teacher can be involved with his students. I had many great tennis games with students. Some were better players than I was, but some I could beat.

I was fortunate to get to know many students very well and I still correspond with a significant number of them. However, all good things must come to an end. As much as I liked the Academy and as much as I enjoyed teaching there, it was time to move on. My entire family lived in New Jersey and I missed not being able to get together with them more often. I was an Easterner and wanted to move back East. An ideal situation would have been if I could somehow pick up Lake Forest Academy and move it somewhere near Princeton, New Jersey. The administrators at Lake Forest Academy were very understanding. I decided that I wanted to find an independent school that was within one hour of either New York or Philadelphia and within two hours of Princeton. I read through *Peterson's Guide* to boarding schools and found twenty schools that looked good in the advertisements and that fit my criteria.

When I first thought of going into teaching I was completely at a loss as to how to search for a position. Having grown up in Princeton, New Jersey, and having learned that Princeton University has a teacher placement office, I decided to take a chance to see if that office would help me find a summer job. It did and right after I had left the office, Harold Corbin, headmaster of Lake Forest Academy, called seeking a mathematics teacher for the fall term. The placement office gave him my name and phone number. Although I had some offers in hand, this man was intriguing and he convinced me to make the trip west to take a look at what he had to offer. Since that frightful plane trip and nine years of teaching had worked out so well, I went back to the Princeton University placement office and again asked if it would help me. The office had a new director, a very understanding man who was willing to go out of his way on my behalf. Mr. Drury said that he received many more requests for math teachers than he could fill with Princeton graduates. I will always be grateful to Mr. Henry Drury for the help and advice he gave me. His name will appear later as I returned for his aid and advice many years afterwards. I was lucky; fourteen of the twenty schools I picked out had openings for mathematics teachers. That January I was set to spend my winter break week at my parent's house in Princeton and do interviews daily with those schools with openings in mathematics.

I had received a letter from Mr. Archibald Montgomery, headmaster of The Hill School, in Pottstown, Pennsylvania. He had instructed me to call him when I arrived in Princeton and we could set a time for a visit and an interview. I called. To my shocked surprise, I received a message that this phone number was no longer in existence. I knew only what I had read in the *Peterson's Guide* about The Hill School. Although that write up was very good, I thought maybe this was not the place that it appeared to be in print. I crossed The Hill off my list as I had thirteen other schools to visit in those six days. I had a very full schedule to meet all these interviews and also to have a little free time to spend with friends on this vacation. One evening, returning home after a long day of interviews, I found a note on the refrigerator door written by my mother. It said that a Mr. Montgomery from The Hill School had called and wanted to talk with me. He left his home phone number and I was to call him no matter what time it was that I got home. So at 11:30 p.m. I called Mr. Montgomery. Somehow he convinced me to squeeze in a visit to The Hill School. He arranged an interview on the morning I was to fly back to Chicago. Pottstown, Pennsylvania, is not exactly on the way from Princeton to the Philadelphia airport. Somehow, Mr. Montgomery was able to convince me that it was. So on the day I left Princeton to fly west again, I stopped by The Hill School.

I had a great interview, I thought. I was particularly impressed with the Director of Studies, Mr. Charles C. Watson. I later learned that he was not overly impressed with me, but he did write to me after the interview that he hoped I would join The Hill School's faculty. The headmaster, who goes by Tad, was impressive also. I also met that day David Mercer, the Athletic Director. I would later learn what a great human being this man is. Even that day, it was very clear that he was everyone's friend. The one lasting comment made to me that day was made by Chuck Watson. Perhaps that alone sold me on the kind of school The Hill School is. He described the teacher contract as a three by five card with but one sentence on it. "I agree to teach at The Hill School for the school year 1970-71." That is all you signed. He stated that at The Hill your word and your handshake have meaning. This impressed me. I was later to learn that Chuck Watson lived that philosophy to the fullest. His word was gospel.

I had my credentials sent to The Hill from the Princeton Placement Office. I had asked Harold Corbin, who had retired the year before, to write one of my recommendations. Although I never saw his recommendation, during several of my interviews, people mentioned that he had written many good things about me. After writing to Harold, I received a letter from him in which he said that his heart sank as he opened my letter because he was quite sure why I was writing. He instructed me not to leave the Academy unless I found the school and the offer that were just too scary to turn down. I am not sure what he meant by the word *scary* in that statement, but I did know that if offered The Hill position I would accept. I also had asked Sid Ainsworth to write a letter on my behalf. Again I never saw that recommendation, but one administrator quoted from it during my interview. "I feel as if I am cutting off my own right arm in writing this letter." Then supposedly he went on to explain what I had accomplished during my nine years at Lake Forest Academy.

I left The Hill School and flew back to Lake Forest. The first thing I did was go to Sid Ainsworth and explain that I had had a great trip and that there were several schools in which I had an interest. There was no question in my mind that The Hill School was number one on my list. Two of the other schools had made offers on the spot. However, I wanted to wait until I had visited all my schools and then determine which would be best for me. Of course, Walter Hoessel, the new headmaster at Lake Forest Academy, made it clear that he would like me to return there. The following day I received a phone call from The Hill School's headmaster, Tad Montgomery. He wanted to let me know that he was putting an offer in the mail and did not want me to make any decisions until I received it. He went over the offer on the phone. I replied I would sign and return immediately the 3x5 card. So my days at Lake Forest Academy were coming to a close and I would be starting over again at The Hill School. I hoped that my first years at The Hill School I would not make the same beginner mistakes I had committed at Brick House School and Lake Forest Academy.

Chapter II

The Hill School Years

I arrived at The Hill School to find that my apartment was not much better than the first one I had lived in at Lake Forest. Just as small, the apartment had a living room, bedroom, bath, and a closet for a kitchen, but it did have a stove. I could not open the oven and the refrigerator doors at the same time. Two people could never be in the kitchen together. The view out the windows was less than dramatic. The only thing that I could see was roof tops.

The building in which I lived was called Middle School. It housed eighty students and five faculty families. My apartment was called Middle School Center and was on the south side of the building. It seemed to have sun shining in all day long. The weather that fall was extremely hot, and the apartment became unbearable. At the end of my first year I asked to move to another apartment. I did not care where on campus the School moved me, but I could not remain at The Hill School if I had to live in that apartment another year. I was offered another apartment in the same building, Middle School 3 West. That was the apartment on the third floor, west end, of Middle School. It was quite an improvement over my previous living conditions. I had a living room, which was about as large as my entire previous apartment. This apartment had a large bedroom, a hallway and a study. The kitchen was small but quite adequate. After a few years the School allowed me to add one of the student rooms onto the apartment. I used it as another study because it provided easy access for students who sought extra help. This added room also gave me an easy way to keep an ear to the noise on the hall.

In those days the salary scale was not very good at The Hill. I took a cut in pay from what I had been making at Lake Forest Academy. In fact, I learned through a secretary friend at the Academy that the headmaster of The Hill School had called to double check on the salary I had told him I earned at the Academy. However, you should know that you do not go into teaching for the money. If your goal is to earn big bucks, do not teach. However, if you want to earn adequate pay and have a fun-filled life and extremely rewarding experiences, teach. Most independent schools do not have published salary scales. Nobody knows what anyone else makes. Salaries are determined between the teacher and the headmaster. My first pay increase at The Hill School was quite disappointing. I had become accustomed to significant pay raises at Lake Forest Academy. I did go in to the headmaster to discuss the situation. I was told that I had come to the School at the top pay it could offer and that it would be unfair to move me ahead of others in the salary scale. I thought then and I still believe that my salary should not depend on what others were making. An individual's salary is between him and the School. I did get my additional raise.

In January 1973 Tad Montgomery announced his resignation as headmaster of The Hill. His replacement was the former Director of Studies, Mr. Charles C. Watson. It is unusual that a new headmaster is appointed from within a school. However, this appointment was enthusiastically endorsed by the entire faculty. The School had been going through some rough times, as were all independent schools in the early seventies. There was much dissatisfaction over the Vietnam War and drug problems were becoming quite common with teenagers. Chuck Watson came to be the perfect person to right the ship and get it sailing on a smooth course. He was an academician so it was quite clear that academically the School would be on solid ground. As Director of Studies he had instituted the independent studies program. A student signing up for an independent study class would work one-on-one with a faculty member. The bulk of the work was done by the student with the faculty member being a guide or mentor.

Mr. Watson called me into his office on one occasion to see if I would be interested in being the College Adviser. I turned him down on the spot. First of all, I had done some administrative work, part time, at Lake Forest Academy, and was extremely disappointed with it. The College Office was the last place I wanted to work. If a student is accepted to the college of choice then the parents are proud because, of course, their son or daughter should be accepted. Every parent knows that his or her child is an "outstanding individual." On the other hand, if the student receives a rejection letter, the parent knows it is the fault of the school and the college adviser and I wanted nothing to do with that scenario.

Fortunately for me Mr. Watson did not take no as a permanent answer to any administrative duties. At another time, he called me in to see if I would be interested in becoming the Assistant Director of Studies. Would I agree to this position if it involved absolutely nothing with college advising? I would teach three sections of mathematics and assist John Woodward, who would be both Director of Studies and College Adviser. Mr. Woodward had been at The Hill School for a number of years and I knew him quite well. He would be an ideal person with whom to work. He had the School's interests at heart, and he was willing to devote much needed time to the students. John Woodward divided up the duties so that I would work with the eighth, ninth and tenth graders and he would take care of the eleventh and twelfth graders since so much of the advising for them would involve colleges.

This arrangement in the Director of Studies Office went well for two years. However, Chuck Watson decided that college advising required someone full time and asked me to take over as Director of Studies and John Woodward to be the College Adviser. I accepted with the proviso that I continue to teach two sections of mathematics. My greatest joy at the School was the classroom experience and a working relationship with the students. I did the Director of Studies job for eleven years. It was challenging and I am glad I did it, but I ended up spending more and more time with parents and faculty and less and less with students. I wanted to return to the classroom and be a full-time teacher. I also was very interested in becoming department chairman. There were several things which I thought a math department should be doing and wanted to institute these ideas. The Hill School policy was that a department chairman was appointed and remained in that position until he retired. The current chairman was not much older than I, and I was not willing to wait that long. I loved The Hill School and wanted to stay on its faculty, but I thought I could do so much more in building a mathematics department. So I submitted a letter of resignation as Director of Studies in which I stated that I wanted to look for a job as a department chairman, but if I were not able to find one, that I would like to return to The Hill as a full-time teacher of mathematics. I went back to Mr. Drury at the Princeton University Placement Office. He helped me find some schools that were looking for chairs of their mathematics departments. I was very fortunate in that several of these schools fulfilled my requirement of being within two hours of Princeton. I did receive two offers so I knew that I might be a department chair somewhere the following year.

Several years earlier I had submitted to the trustees that, like many other schools, The Hill should have a rotating department chairmanship policy. I was called into the Headmaster's Office and given a proposal which ideally fit everything I wanted. Happily Chuck Watson offered me the Chair of the Mathematics Department at The Hill School. The School would institute a rotation policy in that a chairman would be appointed for a five-year period. The chairman could succeed himself once, but after ten years someone else would be appointed. That offer could not have been better. I would get to remain at The Hill and be the mathematics department chairman.

As chairman the first thing I instituted was an honors program. We began that first year with an Algebra I Honors class. The next year we added a Geometry Honors class and then continued to add honors sections each following year. I felt strongly that our top students had not been getting all they could academically because they were mixed in with the average students. We had some students who were very talented in mathematics and deserved to be stretched to their limits. We then took a close look at the textbooks which had been used for so many years at The Hill. Calculators were just then beginning to be used extensively in the classroom but none of our textbooks had any calculator material in them. In addition, we began using some of the national math contests. We gave the monthly contests which were administered to students in specific math courses. One test was given for each specific subject area. Then we administered the big national Mathematics Contest, designed for the most talented students. This test is one of the most difficult tests I have seen and I am quite sure that I could not achieve a perfect score on it myself in the given time limit. Bright students enjoy stretching their minds and seeing what they can do. These tests also show some students that they may not be as accomplished as they think they are. This recognition can be shocking to them, but it is still a worthwhile experience. Fortunately, we had Larry Kelly on our faculty at that time. He is an outstanding teacher, one of the all around best I have known, who also had the ability to inspire students to work hard for these national competitions. Our results were outstanding, mostly through the inspiration of Larry Kelly.

At an independent boarding school every faculty member wears many hats and I was no exception. All faculty members are expected to coach at some level in the athletic program. Athletic

Director, Dave Mercer, always tried to juggle around the assignments to cover all the teams and have the faculty members involved in the areas they most enjoyed. Although I have loved sports my entire life, I have never really enjoyed coaching. In the classroom I knew several different ways to explain a given lesson. If a student did not understand it the first time, I could try another approach. But, coaching was different. I could show an infielder how to approach a ground ball, how to bend his knees, how to stay in front of the ball, and how to be in the best position to make the throw after fielding the ball. If he could not do it after that, all I could do was repeat my instructions over and over again. I found coaching to be quite frustrating. Also I was not serious enough about sports at that level. Winning in my mind was certainly not the most important thing, but I liked the idea that every student was required to take part in athletics all three seasons at The Hill School. There are great learning experiences that can be derived from participation in sports. A student can excel in the classroom and have no solid relationship with the other members of the class. In sports, particularly team sports, that isolation is unlikely. What an athlete does on the field affects the entire team. The togetherness that results from the many hours of practice and all the ups and downs of a season can be learned nowhere else. This relationship among teammates is what I always wanted to emphasize along with good sportsmanship and morals. I am sure that is what Harold Corbin meant when he emphasized teaching the whole student. I also tried to make the game I coached fun. Remember, some members of those teams were there because they were fulfilling a requirement, not because they loved the games. I would try everything so that a day on the field was an enjoyable afternoon. Most importantly though, sports to me were to be learning experiences equal to those the student received in the classroom. As a result, in different years I did coach baseball, basketball, and tennis, all at the eighth and ninth grade levels.

In juniors baseball eighth and ninth graders would practice pickoff plays. One of my favorites was to try the Goodyear Blimp play. As the runner was starting to take his lead off first base, the first baseman would say, "Look, the Goodyear Blimp" as he pointed skyward. At the word "blimp" the pitcher would fire to first to pick off the runner looking skyward. We practiced this play on many occasions, but it worked only once in an actual game.

Once, while I was coaching juniors basketball, we were down at halftime by thirty points. We were completely outclassed but during halftime I devised a play that would keep us in the game. In those days only one referee worked the junior-level games. I had one player designated as the sixth man. After the referee passed our bench and headed down the floor, the sixth man ran onto the floor. Then I sent a player to the scorer's table as a sub to enter the game. When a whistle blew, the substitute ran onto the floor, not waiting for the referee to wave him in. Off the floor came the player the sub had replaced and the sixth man. Thus, at no time were there six men on the floor during a free throw, when it would be obvious that we had too many men in the game. Furthermore, the sixth man was instructed never to shoot. He was there for defensive purposes and rebounding. I did not want him shooting and having the opposing team trying to search out who was supposed to be guarding him or they would realize six players were on the floor. We played a man-to-man defense with our sixth man under the basket for rebounding help. Although we did not win the game, we did play the opposition even in the second half. Near the end of the game a student walked out of the stands and around the court. Coming to the bench, he whispered in my ear, "Mr. Mercer says you have six men on the floor." I nodded my head yes and looked up in the stands until I could make eye contact with Mr. Mercer. I then just smiled and nodded my head yes.

In later years I coached juniors tennis, but I felt very inadequate there. The only training I had was from the two students at Lake Forest Academy. It was not like coaching the game I had played since I was a young kid. However, Mr. Mercer needed someone to cover that sport, so I did it for him.

Another duty was serving as chair of the Scholarship Committee. Every family requesting financial aid had to submit a financial form. The committee reviewed every form and then tried to divide the limited resources fairly. The Hill School did not have unlimited scholarship funds and therefore there never was enough money to fulfill everyone's need. Being fair to every family usually meant that no one was given the amount of aid that we really would have liked to give. One of my worst experiences with a family occurred while I wore this hat. The parents of a student already enrolled applied for financial aid. No matter what criteria were used, there was no way this family could qualify for any financial support. I wrote what I thought was an appropriate letter stating that we were unable to fulfill their request because others had greater financial needs. I received a letter from the father stating that I was the worst person he had ever known and that he hoped he never saw me when he visited the campus to see his son. I did not reply to him and I never said a thing to his son, but I often wondered whether the boy knew what his father had done. This incident came back to me often because after the son graduated he returned to campus for alumni functions. He always stopped at my house for a visit when he was on campus.

Serving as a dormitory parent is a very important and rewarding duty for a boarding school teacher. You get to know the students under your charge very well. You should; you live with them for thirty weeks a year. In the spring of my first year at The Hill School I was named head of Middle School. So with four other faculty members to help me, I was in charge of eighty students, twenty-two of whom lived on my hall. Dorm parenting

20

is an extremely time consuming job if done correctly, but well worth the effort. The Hill School was always quite generous in providing money for hallfeeds. Thus, a dorm parent could have those students who lived on his hall into his apartment and serve them something to eat. The hallfeeds varied from some hallmasters' full course meals once or twice a year to my usual biweekly pizzas, or sandwiches, or ice cream sundaes.

Each student at The Hill School is assigned a faculty adviser, who keeps very close tabs on that student. The adviser is the first person parents contact if there is a problem or the parents just need information. A teacher may become close to those students on his or her hall, but the relationship between an adviser and advisee is astounding. The young men and young women I had as advisees, I will always remember. I was able to cry with them over their miseries and congratulate them in their triumphs. Perhaps it is because I never had children of my own that I always felt the relationship I had with advisees would be very similar to that of a father to a son or daughter. I could write another entire book just with experiences I had with advisees. These young men and women come back often to visit and I cherish our relationships to this day.

The Hill School has a very good student workjob program. It is felt that every student should be required to work for twenty to thirty minutes per day for The Hill School. In this way students help to pay for their own education. Therefore every student is assigned a workjob. Some students work in the major offices on campus, where they help with filing or delivering messages. Two students are assigned to raise and lower the flag every day for the entire year. The majority of the students have workjobs in the dining hall where they set tables and clean up after meals. The Hill still has eight rather formal sit down family style meals a week. The rest of the meals are buffet style. For the buffet meals there is no table-setting, but there still must be a clean-up crew after the meal. For many years I was in charge of the students assigned to dining hall workjobs. A very stringent hierarchy system assigns new students to the menial tasks and older students as row captains to oversee that the jobs are done properly. A head dining room captain, the one student in charge of the whole operation, has a very tough job and great responsibility. I always claimed that if he or she did not do the job we might not have the next meal. The first two or three weeks of the school year were tedious in the dining room as each student learns his or her job. I always tried to set the workjobs up so that the students ran the show. After the first few weeks I was able to sit in the corner and just watch the captains run the show. For the most part we had great captains and the program ran very well. I was able to get to know many students by seeing them in the dining hall every day doing their workjobs. I roamed around and talked with the students as they worked and I learned about their successes and problems in certain classes. If a student missed his workjob for some reason, often I would write a short note to him. It was as if the note were from the actual table he was supposed to set. The table would make it clear that it was upset at having a substitute table setter as this stranger did not treat the table properly.

There is no question that a faculty member's major duty is classroom teaching. However, I will save those experiences for another chapter. I want to devote an entire chapter to such a major responsibility.

Chapter III

Autographs

Living in Pottstown, Pennsylvania meant that I was near my parents, who were still living in Princeton, New Jersey. This was quite fortunate as I could make a quick trip in less than one and one-half hours to see them. At age 82 my mother fell and broke a leg and an arm. This accident completely demoralized her. In her prime, which lasted until the time of her fall, she was a very active woman. She loved people and she was always busy. After the fall, as she recuperated in the hospital and then later in the nursing home, I was sad to see she had lost interest in everything.

One night early that fall I was thinking of possible gifts I might get her for Christmas. No matter what I thought of, I decided that she would not like it. What would bring her just a little enjoyment? Only two things would bring a smile and friendly chat from her. She enjoyed receiving mail and she liked having visitors. Although I knew it would be impossible to get her many visitors, I thought I could get her some good, interesting mail. I would write to people she admired, describe the situation, and ask each of them to send her a Christmas card. I walked around campus the next few days stopping everyone I saw, asking if you could get a Christmas card from anyone in the world, whom would you want it to be? I finally made a list of 100 people for whom I thought she would enjoy receiving cards and, I was able to find current addresses. She received 59 responses from this project and I received 14 responses, letting me know that they thought this was a great idea.

My mother did not receive a Christmas card from Margaret Thatcher, but she did receive an autographed photo. From the White House came a card inscribed "The President and Mrs.

Reagan extend to you their best wishes for a joyous Christmas and a peaceful New Year." Joe Garagiola added to his Christmas card "Just a note to let you know we are thinking about you." He also included twelve baseball cards, all of himself. On a separate piece of stationery he wrote "Thought I would put in a couple extra cards in case you have the chance to make some trades. If not, you can be my press agent and hand some out for me." Millicent Fenwick of New Jersey sent an official Christmas card from the House of Representatives. In addition, she hand wrote a short note to me. "The Christmas card to your mother is going out in this same mail. I only wish every mother had such a good son!" Willie Mays, known for not being the most agreeable person on this earth, signed a Christmas card and also included an autographed baseball card. Betty and Jerry Ford sent their personal Christmas card with photos of their grandchildren. I had not been able to find an address for the Fords so I had sent my request to the White House with a note on the envelope, "Please forward." It worked!

Growing up, Bobby Thomson was my favorite baseball player so I just had to try to get him to respond. He included the following message on his card. "Through the love and warmth of your son Willis, I am given the opportunity to wish you a Merry Holiday Season. I hope you get a lot of Christmas cards this Christmas and your spirits remain high through the year." Lawrence Welk included an autographed record jacket. John Cameron Swayze included on his card "I am delighted to have this chance to wish you all the best, not just at the Yuletide but all through the New Year." In addition, he replied to me with the simple note, "Glad to do it. Merry Christmas."

Rose Kennedy signed a card that included two photos, one each of her sons, Jack and Bobby. Barry Goldwater wrote a letter from the United States Senate. "A little bird has told me you are presently living in the Princeton Nursing Home, and I wanted to wish you a very Merry Christmas. I hope Santa will be good to you and that you will be blessed with good health and happiness in the coming year." Walter Cronkite included a letter. "You have a wonderful son with whom I have corresponded and I just wanted to send you my very best wishes for happiness and health during this holiday season, and a very Happy New Year!" Art Linkletter wrote "with" and drew a heart with an arrow through it. George Burns, who at the time was probably only a couple years older than my mother, added the following to his card. "Stay happy kid!"

John Lindsay added "with all good wishes to a very good person who must be proud of her son Willis." David Brinkley sent my mother a card and added a best wishes note. He also added a note to my letter and sent it back to me. It has to be the shortest message I have ever received, but it explained everything from a man who obviously was very, very busy. "WJP--done--DB"

My mother also received cards that year from Skitch Henderson, Louis Nizer, James Garner, Ted Kennedy, Morley Safer, Norman Vincent Peale, Merv Griffin, Andy Williams, Elizabeth Taylor, Bess Myerson, John Chancellor, Roy Rogers and Dale Evans, Monty Hall, Robert Young, James Stewart, Carmella and Frank Rizzo, Lady Bird Johnson, Leonard Bernstein, Bob Hope, Red Skelton, Mitch Miller, Robert Stack, John Updike, Gene Kelly, Bette Davis, Harriet Nelson, Danny Thomas, Oral Roberts, Michael Landon, Paul Harvey, John Connally, Yogi Berra, Perry Como, Betty White, Johnny Carson and Liberace.

The final card arrived on February 20th. It was a Christmas card to which the following note was added. "Happy Valentines' Day too!--Love--Jonathan Winters."

This project turned out to be better than I could have dreamed. Most of the nurses at the nursing home visited my mother every day just to look at the cards she had recently received. Also as a Christmas gift, I bought a nice album to store all her cards. On Christmas day we spent the entire afternoon filling that album. Of course she was not able to do any of the work because of her broken arm. However, she would tell me which ones to put in and where. On several occasions she changed her mind because she did not think those two people fit together on the same page. Also we were able to talk about all these great people who took the time to send her a card. It made a huge difference in her attitude toward that Christmas season.

When my friends look through this autograph album today, I get kidded about the letter I wrote to these celebrities, requesting them to send Christmas cards to my mother. It may be a little misleading, but not by much.

"My mother has been confined to a nursing home for some time. Her only enjoyment seems to be having visitors and receiving mail. Since you are the one person she always admired, I would like to ask if you would send her a Christmas card. It would most certainly brighten her spirits during the holiday season. I thank you for your kindness."

Since so many people were kind enough to reply to a mail request, I thought it would be interesting to start an autograph collection of my own. I had previously collected some autographs of baseball players, but always in person. I wanted to get each autograph for my new collection on something personal. The only thing I could think of that anyone would willingly give me was a business card. I began by writing to former major league baseball players. I thought it would be interesting to have their autographs on the business cards of their new professions, those they entered after their baseball careers were over. Not a large number of ball players had business cards so I expanded to all sports. Later I expanded further to all walks of life. Many people did not have business cards, but they would usually sign something for my collection. I received autographed books from Dick Vitale, Wally "Famous" Amos, and James Earl Ray. Many people sent autographed photos as well.

Often I am asked which autograph is my favorite. There is no question in my mind because I never really expected to get a response and I was quite sure that the person would not have a business card. However, I received a nice looking prayer card and at the bottom was the autograph of Mother Teresa. This is my favorite.

Chapter IIII

The Classroom

The goal of any teacher should be to create a stimulating classroom atmosphere so that each student looks forward to attending that class every day. Both the appearance of the classroom and the way class is handled are important.

One wall of my classroom was a bulletin board on which I put up two hundred and fifty, four-by-six photographs. About every three weeks I replaced the set with new photographs, mostly sports photos of the students who were in my classes. I did mix in some other students and also campus scenes. Many students would just stop in to visit and to look at the photos. Those photos they liked best I enlarged and hung in the main hall. The Hill School permitted me to frame photos and decorate the mathematics end of the Academic Center with these enlargements. I would enlarge the photos to an eleven-by-fourteen format and then matte and frame them in sixteen-by-twenty frames. There were usually over eighty photographs hanging in the hallway at any given time. The enlarged photos were for sale and many students and parents did buy them. Although I cannot say I made a profit, I can say that the additional income allowed me to take many more photos.

On the side chalk board I always made a list of the autographed business cards I had received most recently. As students came into the room, they would check the names on the board and then try to figure out who the people were. Many of the students would just guess that they were former baseball players since most autographs in my collection are from major leaguers. It was interesting to hear their responses and guesses. I was always amazed when a student identified someone I never thought he or she would know. It was also interesting that the younger students often came up with answers when the older students were stumped. If no one came up with the answer, I would pass the business card around the classroom. Usually this was a giveaway, but sometimes the business card gave no clue to the person's identity.

Following are some of my more famous (or in some cases infamous) autographed business cards. My Heisman trophy winners cards include Jay Berwanger, Gary J. Beban, Joe Bellino, Angelo Bertelli, Tony Canadeo, John R. Cappelletti, John David Crow, Peter M. Dawkins, Clinton E. Frank, Archie Griffin, Tom Harmon, Paul Hornung, Leslie Horvath, John Huarte, Vic Janowicz, Richard W. Kazmaier, Jr., Larry Kelley, John Lattner, O.J. Simpson, Steve Spurrier, Roger T. Staubach, Pat Sullivan, and Doak Walker. I have received cards from other football people: George Allen, Raymond Berry, Rocky Bleier, Terry Bradshaw, Dick Butkus, Weeb Ewbank, Joe Gibbs, Paul Giel, Gerry Glanville, Otto Graham, Rosey Grier, Louis R. Groza, Lou Holtz, Lamar Hunt, Jimmy Johnson, Charlie "Choo Choo" Justice, Tom Landry, Steve Largent, Edward W. LeBaron, Jr., Sid Luckman, Wellington T. Mara, Arthur B. Modell, Bill Parcells, Ara Parseghian, Joe Paterno, Joseph Robbie, Kyle Rote, Pete Rozelle, Buddy Ryan, Texas E. Schramm, Donald F. Shula, Bart Starr, Francis A. Tarkenton, Joseph R. Theismann, John Unitas, and Stephen B. Verbit. Baseball people who have autographed cards include Henry L. Aaron, Dusty Baker, Howie Bedell, David J. Bresnahan, Jim Bunning, Bob Cain, Larry Doby, Dave Dravecky, Gerry Dvorak, Carl D. Erskine, Mark S. Fidrych, Charles O. Finley, Joe Garagiola, Harvey Haddix, Jim "Catfish" Hunter, Bowie K. Kuhn, Johnny Mize, Don Newcombe, Ed O'Brien, John O'Brien, John "Blue Moon" Odom, Phil Rizzuto, Marge Schott, Enos Slaughter, Carl E. Stotz, Robert B. Thomson, and Jean R. Yawkey. People who have gained their fame from basketball and have sent me business cards include Arnold "Red" Auerbach, Rick Barry, Elgin Baylor, Bob Cousy, Thomas J. Gola, Gail Goodrich, John Havlicek, Dan Issel, Earvin Johnson, Mike Krzyzewski, Jeff Mullins, Willis Reed, Bill Sharman, Earl Strom, Jim Valvano, Jr., Jerry West, and John R. Wooden,

Recently astronauts have not been very good about answering autograph requests. My guess is that they get so many requests that it is impossible to respond to every one. However, I have received autographed business cards from Neil A. Armstrong, Frank Borman, Gordon Cooper, John Glenn, Wally Schirra, and Alan B. Shepard, Jr. Politicians, for obvious reasons, are very good signers. However, it is hard to tell whether some of the business cards are signed with an autopen. I have received business cards from Spiro T. Agnew, Marion Barry, Jr., Barbara Bush, George Bush, Jimmy Carter, Rosalyn Carter, Bill Clinton, Hillary Rodham Clinton, Bob Dole, Geraldine Ferraro, Gerald R. Ford, W. Wilson Goode, Jr., Alexander M. Haig, Jr., S.I. Hayakawa, Jesse Helms, Daniel K. Inouye, Edward M. Kennedy, Henry A. Kissinger, Walter F. Mondale, Edmond S. Muskie, Thomas P. O'Neill, Jr., Ronald Reagan, James Bond Stockdale, Strom Thurmond, and George C. Wallace. The following business leaders have also been gracious enough to answer my requests for business cards: Perry Bass, Henry W. Bloch, August A. Bush, III, Roy E. Disney, Michael D. Eisner, Malcolm S. Forbes, Julio and Ernest Gallo, William H. Gates, Lee A. Iacocca, Franklin P. Perdue, Paul Smucker, Frank Steele, Peggy Steele, R. David Thomas, and Donald J. Trump. I should have collected more educators, but I do have cards from William Ayers, Penelope H. Dunham, William Dunham, C.C.F. Gachet, Timothy Leary, Matthew B. Ralston, Carter P. Reese, and Thomas G. Ruth.

Not fitting into any of the above categories, but certainly notable enough to be included in this list are Mario Andretti, F.

Lee Bailey, Carmen Basilio, Shirley Temple Black, Brian Boitano, Art Buchwald, William L. Calley, William S. Carpenter, Jr., Don Carter, Florence Chadwick, Joseph J. Cicippio, James D. Craig, Michael E. DeBakey, Hugh Downs, Russell Drowne, Angelo Dundee, John Ehrlichman, Julie Nixon Eisenhower, Kenneth J. Feld, Milton Friedman, Gene Fulmer, Dan Gable, Philip G. Gallagher, Frank Gasparro, Larry Holmes, Bobby Hull, Bruce Jenner, William Joel, Jack S. Kilby, Billie Jean King, Wade Kirby, Kreskin, Rod Laver, James R. Leavelle, Jerry Lewis, G. Gordon Liddy, Marty Liquori, John Longden, Phil and Steve Mahre, Henry Mancini, Bob Mathias, Willie Mosconi, Sterling Moss, Byron Nelson, Justin W. Newton, Manuel Antonio Noriega, Oliver L. North, Alex Olmedo, Arnold Palmer, Fess Parker, Floyd Patterson, Roger S. Penske, Dick Perez, George A. Plimpton, John Marlan Poindexter, C.G. Rebozo, Norman Schwartzkopf, Bill Shoemaker, Frank Shorter, Pam Shriver, Sam Snead, Lt. Jay Spencer, Oliver Stone, Lee B. Trevino, R. E. Turner, Al Unser, Paul A. Volcker, Ellsworth Vines, Dr. Ruth Westheimer, William Childs Westmoreland, Jeff Whitmore, and Rev. George J. Willis, Jr.

The reader may want to try to count the number of names on the list that he or she recognizes and can identify. For several of these cards I would like to add a comment or two. I have written to a number of these people on more than one occasion. Politicians and athletes particularly who move around and get different business cards for their new jobs make particularly interesting contrasts. For example, I have a George Bush Sr. card as vice-president, another as president, and yet another as an ex-president business card, with nothing more than his name on it. Carl Stotz's card has a photo of a little league game on the front. He signed the back. The reverse of Howie Bedell's card has a nice expression of "The Winning Attitude..." Bill Carpenter wrote a nice letter to go with his card. He evidently thought I was a student. He explained that collecting business cards was a very good hobby, but the most important thing, he warned, is academics; do not let the hobby interfere with your learning. Although I was not a student this was great advice for my students, especially coming from the lonesome end. Weeb Ewbank had the following on his card. Under his name is "retired." In the four corners are the words, "No Money" – "No Business" – "No Address" – "No Phone." Julie Nixon Eisenhower has a blank card. She wrote on it "with best wishes" and then signed her full name. Harvey Haddix also signed a blank card. Julio and Ernest Gallo each sent an individual card. I got Archie Griffin's card by walking into his office one day at Ohio State University and asking him for an autographed business card. He also signed a book and a baseball for me that day. (I also collect autographs on baseballs.) Larry Holmes' card has his photo on it. Larry Kelley's card has a gold embossed Heisman Trophy on it. Wade Kirby's card is signed Wade Howard, his stage name. John Longden had his telephone number on it and then directly under the number was 6 A.M. to 9 A.M. I assume at other times he was with his horses.

Although I do not pay for autographs sent through the mail, I have received a number of replies each stating an amount of money that I must send with my request for a signature. I did relent once. Johnny Mize asked that I send two dollars and that I make out the check to the Boy Scouts of America. Floyd Patterson added the note "To my friend Willis." Matt Ralston had obviously used the card as a scratch pad as he had written someone else's e-mail address on the back. Frank and Peggy Steele each sent a card. Jay Spencer wrote on his card "Thanks for opening my eyes to all the possibilities," and signed it simply Jay. John Wooden's card has his pyramid of success on the back.

Also in my classroom I kept a stack of eight-by-ten signed photos I had received through the mail. These are ones that either I really liked or I thought the students would enjoy seeing as well: Kareem Abdul-Jabbar, Richie Ashburn, Gene Autry, Yogi Berra, Jay Berwanger, Earl Campbell, Phyllis Diller, Dom

32

DiMaggio, Chris Evert, Farrah Fawcett, Bob Feller, Al Gore, Pete Gray, Jesse Jackson, Ralph Kiner, Bob Knight, Elle Macpherson, Mickey Mantle, Willie Mays, Mary Ann Mobley, Clayton Moore, Stan Musial, Jack Nicklaus, Walter Payton, Pele, Dan Quayle, Sally Ride, Kenny Rogers, Mike Schmidt, Tom Seaver, Buffalo Bob Smith, Warren Spahn, Cheryl Tiegs, Arnold Tucker, Johnny Unitas, Mickey Vernon, Barbara Walters, Lawrence Welk, Vanna White, and Oprah Winfrey.

I also have two photos signed by Chuck Connors, one as a first baseman for the Chicago Cubs and the other one in his Boston Celtics uniform. I never got a signed photo of him as The Rifleman. I also have a photo of both Bobby and Brett Hull signed by each of them. There is a nice shot of Stevie Wonder signed with a thumb print. I also have a photo of Hank Aaron and Eddie Mathews in Atlanta Braves uniforms and signed by each. Because I am a long-time New York Giants football fan, one of my favorite photos is Andy Robustelli, Rosey Grier, Dick Modzelewski, and Ken Katcavage. That photo I sent to each of them and finally got it signed by all of them. Hillary Rodham Clinton included, along with her business card, a photo of Socks, complete with paw print. Finally, I have a photo of the Oak Ridge Boys, signed by all four members.

My classroom included such things as a giant-size baseball and a lamp taken from the old Middle School literally minutes before that part of the building was demolished. A sign which all parents really enjoyed when they visited read "Hire a teenager while they still know everything." Some of the students thought it was cruel to keep that sign in the room.

I would often make bets with students. The bet was always one which I knew I would win. The wager was always one can of soda. I kept my winnings lined up in the front of the room so that I could razz the poor losers for the remainder of the school year. Every year I made two statements and I always had someone call me on them, willing to make a wager. "I can give you the score of the Super Bowl Game, right now, before it ever begins." Also, "I can jump higher than this desk!" (Answers appear in Appendix A.)

I usually saved space on the side board for what I called the letters home. If a student sneaked into breakfast late and thought no one had seen him, when he arrived to class he would be shocked. The letter that day would be

"Dear Mom and Dad,
I was late to breakfast today, but nobody noticed."

Sometimes as a student entered the classroom and I noticed he was not wearing socks, I would add the following letter.

"Dear Mom and Dad,
I was in a hurry today and did not put on socks. Nobody noticed."

Walking up to campus one morning at 6:00 a.m., I noticed that the flag was still up. The young man whose work job it was to raise and lower the flag was in my class. I could not resist putting up a letter.

"Dear Peter,
I noticed the flag already up very early this morning. Congratulations on putting it up well before breakfast."

Part way through the year I had many students come to tell me things about their friends so that I could put them in letters home. On those days when I did not have a letter about one of my students, I just made one up that I claimed a student could write home. Here are a few of my favorites.

"Dear Mom and Dad,
How much time do I have left before I get interested in girls?"

"Dear Mom and Dad,
My teacher put an advertisement in the paper for his lost dog. That's silly--the dog can't read. "

"Dear Mom and Dad,
I told my teacher I was going to run away. He gave me $5 and wished me good luck."

"Dear Mom and Dad,
I checked out the two biggest books in the library today. Now I can reach my mailbox."

"Dear Mom and Dad,
Mr. Walbridge's (the biology teacher's) lesson today was not as good as your stork story."

"Dear Mom and Dad,
My English teacher claims I need to improve my punctuation. That's unfair; I have never been late to that class."

"Dear Mom and Dad,
It was nice to see you on Parents' Weekend. Sorry about the pictures on the wall. I don't know why my roommate hung them on my side of the room."

"Dear Mom and Dad,
I could never figure out my position on the football team. At the Awards Program coach said I was the greatest drawback on the team."

"Dear Mom and Dad,
   All my teachers are complaining how their investments have shrunk in the last few months. I don't understand--my baseball cards have doubled in value."

"Dear Mom and Dad,
   Why does my teacher think I was born on April 1?"

"Dear Mom and Dad,
   My teacher claims he saw my picture in a Charles Atlas advertisement. My picture was the first one in the ad."

"Dear Mom and Dad,
   My teacher claims I am an underachiever, but the truth is that he is an over-demander."

"Dear Mom and Dad,
   Why did the kids in my class laugh at me when I told them Santa ate the cookies I left for him?"

"Dear Mom and Dad,
   My teacher is wasteful. He never uses the other side of his transparencies."

"Dear Mom and Dad,
   Thanks for telling me my English teacher has a beautiful vocabulary. I am going to ask if I can see it."

"Dear Grandma,
   Did you ever have any kids?"

   Every letter and comment brought out some kind of response or reaction from the students. Many former students stopped to visit daily, just to look at photos and read whatever was written on the board.

I also kept a backwards clock in the classroom. That clock was numbered backwards and the hands moved around backwards. It was the only clock in the room. Students had to learn how to read that clock if they wanted to see how much time they had left to work on a test. I always claimed this clock was correct and all the other clocks were backwards. My only form of proof was to look at the numbers on the bottom of the clock. My clock had the numbers 4, 5, 6, 7, 8 reading from left to right. These numbers on most clocks are backwards.

Students were always in a good mood as they entered the classroom. The few minutes before the bell rang, was a constructive learning time. It was not always mathematics they learned, but certainly the students benefited from additional knowledge and personal interaction. They also learned that a teacher had a life outside the classroom. They were interested in my hobbies and experiences. Many students requested addresses of people from whom I had received autographs so they too could begin a collection.

Chapter V

Teaching

I began each year as all teachers do the first day by taking roll. I read only the last names from my roster and each student responded with the name he or she wanted me to use. Even if I had had the student before or if it was someone I knew very well, he or she had to answer in this fashion. I taught a number of students who, during their days in secondary school, changed the names they preferred to be called. The extreme was a young man I had three times. His first year he went by the name William, the next year, Will, and finally Bill. Another student was Josh one year; the very next year he became Pokey. This method of calling the roll made sure that students felt comfortable with the names used in class. It also gave students the chance to change names. Several students chose to go by different names than they had been called at home. Whether they were shredding embarrassing nicknames or creating new identities the decision was theirs. Always a shock to the parents when they received the first set of written reports! However, I believed that in every case, parents were more than willing to go along with their son's or daughter's decision. I was honored to have the National Council of Teachers of Mathematics include my system of first-day roll call in their kit for new mathematics teachers.

Algebra I was always a difficult course to teach early in the year. There were a number of students who had learned a little bit of algebra in their previous schools. It was also the case at The Hill School that class was a mixture of second and third formers. The School uses the British system of forms whereby the second form is eighth grade and third form is ninth grade. I always gave a test on the first day of Algebra I. I informed the students that I

38

started the year with a test and I ended the year with a test. They did not need to worry though; this test was to give me an indication of how much algebra each student already knew. It would in no way affect their grades. The test usually had ten easy questions from ten different areas of Algebra I. Almost every student would get a few correct. Some students would get six or seven correct. No student ever had enough algebra to get as many as eight correct. It is true that I used this test to give me an indication of how much algebra the students already knew. But, I had an ulterior motive. The test I gave them on the last day was really the same test. On that day I told them that since I did not think they really wanted to take a test, I had someone else take the test for them. I then handed out the test with the name Joe Student on the paper. Together we graded Joe's test. Of course every answer was wrong. I had chosen the most ridiculous or funny answers from the first day's papers to be the answers Joe Student had written down. I called on each student to solve one problem and explain what Joe had done wrong. That student had to determine whether Joe deserved any part credit. Of course, the students were not chosen at random although I tried to make it look as though I did. The student chosen to explain a problem was the one whose work I had picked from the test on the first day. At this point, the students knew their algebra and knew the answers were ridiculous. They laughed at the mistakes and made wild comments. Of course I encouraged these comments about the problems they were discussing. One year a young man said to me that the mistake was so ridiculous that nobody could ever have written it down! I must have made up that answer! After we had decided what grade Joe had earned, I told them that I had another student take this same test. I wanted them to take a look at this other student's work. Then I passed out each student's first day test. Fortunately my students have always been able to laugh at themselves. Students have come back to me years later and asked if I still did that last-day test.

My next-to-last year of teaching, during the second week of school I had a young man enter my classroom as the bell was ringing to begin class. All of my students were at their seats, waiting for the class to begin. This new youngster walked up to the front of the room with a great big smile and said, "I am Patrick Lundquist. I was in Geometry Honors and have been told that I was to report to your class because my algebra is not strong enough to do geometry." He then put his hand out to shake hands and asked where he should sit. I should add that he emphasized the word *honors*, that he was coming from an *honors* class into this regular section of Algebra I. I was amazed such a young man --and a new student at that--could be so poised and so outgoing. My immediate thought was either this was an exceptional youngster or this was going to be a long, long year.

Patrick was a day student and therefore did not have a seat in the dining hall for the evening meals. Sometimes day students stay for dinner because they have a meeting at night. One evening Patrick came walking into the dining room looking for a seat. I suggested that he sit at my table. It so happened that I was sitting at the varsity girls squash table. I thought it would be interesting to see how this new boy third former would interact with all-- mostly sixth form--girls. You would think he had known them for years. He joined right in the conversation as if he had been a regular member of the table. At one point he asked if he could go out to the kitchen to get some cereal. Although I did not care for students' eating cereal at the evening meal, I always allowed them since it was School policy. He came back and ate his cereal but still had much milk left in the bottom of the bowl. He picked up the bowl, ready to drink the milk. I let out a yell, stating that we did not do that at this table. He looked in wonderment and simply asked, "Do I just waste the milk?" I told him he could use a spoon. After a few spoonfuls he looked at me and explained, "I never drank milk with a spoon before."

One of the first topics in Algebra I was basic operations with signed numbers. After we had reviewed the rules I asked my students if they wanted to see the real reason why a negative times a negative equals a positive. I explained they would never see this reasoning in a textbook, but it is the real reason. Of course they wanted to see it.

The Hill School's big rival in sports is Lawrenceville School. I asked the class to answer six questions, and every answer had to be "positive" or "negative."

1. What is Hill School?   +

2. What is Lawrenceville?   -

3. What is it to win?   +

4. What is it to lose?   -

5. What is it if it is good?   +

6. What is it if it is bad?   -

Then I made a chart from these answers.

+ Hill	- Lawrenceville
+ Win	- Lose
---------	--------------------
+ Good	- Bad

Then there were four more questions.

1. Is it good or bad if Hill wins?   +

2. Is it good or bad if Hill loses?   -

41

3. Is it good or bad if Lawrenceville wins?   -

4.   Is it good or bad if Lawrenceville loses?   +

Since the answer to question number four had to be "good," since both Lawrenceville and a loss are "negative," it is easy to see why a negative times a negative was positive.

I love all kinds of mnemonic devices. I always asked someone to spell *mnemonic*. Seldom could an Algebra I student get even the first letter correct. I left the problem unanswered for a day and then would ask whether anyone had looked it up. There would always be two or three students who had researched it.

All students learn in grade school that pi equals 3.14. When the number pi first appeared in Algebra I, I would insist the equality is not fact. Pi is approximately equal to 3.14. Actually the number pi has its decimal places go on forever, the same as some fractions, but the sequence of decimal places for pi never repeat. Wouldn't it be nice to have some easy way to remember the digits out to more than two places? And so I began.

Everyone knows that pi begins with 3.14

What number have you written after 14 more than any other number?

So then we had 3.1415

Besides 15 what number have you written after 14 more than any other number? There were always all kinds of guesses, but nobody guessed it correctly. The answer is 92! I then had all these faces with big question marks in their expressions. So I

42

asked, have you not written the number 1492? They laughed and always remembered that pi is approximately equal to 3.141592.

I usually picked on two or three specific students in each class. I was very careful to choose kids who could take a joke and I had to make sure I could also. It is interesting that I usually picked on students who were fairly good in math, but usually not the very top students. I think these students had a great advantage because they learned from mistakes. I usually called on them when the obvious answer was not correct. They learned to look at things in a different light and could eventually work outside the box. At the end of an Algebra I year, I had several girls come to see if we could have the same students together for Algebra II class. I told them that because of scheduling it was quite unlikely. They asked if just the three of them could stay together along with Brandon and Brendan. I asked why. They answered that they felt safe with them in the class because I always pick on those two boys.

Now back to pi. I asked one of these picked-on students whether he minded if I told the class how he remembered the value of pi. He always agreed. So I wrote on the board "How I want a drink, alcoholic of course." It did not take long for a student to ask how that would help.

I used a similar mnemonic device for the approximate value of the square root of three. Seldom could anyone give me an approximation. I informed my students that it is easy to remember because it is the same year that George Washington was born. Nobody knew the year George Washington was born. I usually let them look it up that night. If nobody said anything the next day, I asked for Washington's Birthday as a bonus on the next test. Afterwards everyone knew a good approximation to the square root of three is 1.732. After they had learned the year, I asked the students for the month and day on which George Washington was born. I warned them that some encyclopedias may have the wrong answer. The next day I usually had a student willing to explain what he or she had found. George Washington was born on February 11, 1732. The calendar was changed in 1752 and left out eleven days. Thus, George Washington celebrated his 21st birthday on February 22, 1753.

Early in Algebra I we studied the three axioms of equality.

Reflexive $x = x$

Symmetric   if $x = y$ then $y = x$

Transitive   if $x = y$ and $y = z$ then $x = z$

The students always seemed to understand them as they make perfect sense. The kids were used to working with equality and were familiar with its use. To make a point I then asked if "less than" satisfies all three axioms? Finally I asked about a non-mathematical term "is a brother of." Would this term satisfy all three axioms?

Reflexive   x is a brother of x

Symmetric   if x is a brother of y then y is a brother of x

Transitive   if x is a brother of y and y is a brother of z then x is a brother of z

Most students missed the symmetric property because they didn't stop to think that y may be a girl. Also they missed the transitive property because it follows the question before it. In this case it makes no difference whether z is a girl. The point was to be sure that the students looked at each case separately.

Before entering Algebra I, students had usually studied order of operations, but some students needed to be reminded. Do multiplication and division first and then addition and subtraction.

$$\text{Thus } 3 + 4 \times 5 = 23$$

I wrote this property on the board as (MD)(AS) and then gave

$$8 + 8 \div 4 \cdot 2$$

Most students came up with 9. Then I explained why I wrote the expression as I had. It is multiplication and division first, but if they both appear do them in the order in which they appear. I always liked to explain that there are no proofs for rules like this. The rule is just accepted by all mathematicians so that we cannot get two different answers for a single problem. They would see many examples of mathematical conventions accepted for convenience. (Example: The algebraic expression 3x is always written with the numerical part first.)

My first year teaching, I used the mnemonic device "My Dear Aunt Sally" for remembering the order of operations. However, one student got confused on the next test and all he could think of was "Aunt Sally makes doughnuts." So he proceeded to do addition and subtraction first. Thus, he missed every problem. Then I switched to "Pity my days at school." The *P* stands for parenthesis. There was always a student who claimed previous instructors had taught "Please excuse my dear aunt Sally" in which the *e* stands for exponents. I do not like that device because an exponent is just repeated multiplication and we already have multiplication in the equation.

Word problems are always fun, but students have learned to think they were difficult. I especially like problems with unexpected answers.

When we began motion problems, I wrote the first one on the board and let the class vote for the correct answer. "If a man drives to work at 60 miles per hour and returns home over the same road at a rate of 40 miles per hour, what is his average rate for the round trip?"

(A) 40 mph (B) 48 mph (C) 50 mph (D) 52 mph (E) 60 mph

Seldom would anyone vote for A or E. Most votes usually went to C. On several occasions students voted for B or D. When I asked why, they explained that it was obviously not A or E and that I would not have asked them the question if the answer were C so it had to be B or D. That was sound thinking. As in all math textbooks there is always work left for the reader. I will leave this problem for you to solve.

Problems involving average are usually interesting. The College Entrance Examination Board usually has problems dealing with average rate of speed or baseball average on the achievement tests.

Assume that on the last day of the season two baseball players are tied with identical batting averages of .400, same number of hits and same number of at bats. Furthermore, assume the batters have at least 400 at bats. In the first game of a doubleheader that last day, Frank First goes 5 for 6 and then sits out the second game of the double header. Sam Second plays both games and finishes with 8 for 12. Which batter wins the batting championship? The catch is that you cannot average averages.

I always began mixture problems with "A person has one quart of an alcohol solution which is 90% alcohol. How much water must he added to dilute it to a 50% solution." I insisted all the students needed to know was what we start with plus what we add equals what we end up with.

We started with 1 quart which is 90% alcohol solution

We added x quarts.

We ended up with 1+x quarts of 50% alcohol solution

What percent had we added? I received all kinds of answers, but finally someone realized that it was zero. I explained "Of course there is no alcohol in water; if there were; Johnny would be out by the drinking fountain all day long."

I used the names of my current students for the names in my word problems on tests. A problem could read, "If Lisa, working alone, could mow the Quadrangle in 30 minutes and, Patrick alone could mow that same Quadrangle in 6 hours, how long would it take to mow the Quadrangle, if they worked together? You may assume that Ross is not around to get in their way."

On a dreary Friday I gave the following question and explained that on Monday each student had to be able to persuade the rest of the class that his or her answer was correct. Just getting the correct answer was not enough. "There is a car on the edge of a 100 foot cliff. At the foot of the cliff is a lake with a boat in the middle of the lake. There is a taut tow line attached to the boat and the car. As the cars moves forward to tow the boat toward the cliff, which moves faster, the car, the boat, or are they the same?" Students argued for each of the three choices. One year I gave this question to three classes, an Algebra I class, a Geometry class, and a Calculus class. On Monday I had students volunteer to demonstrate the correct answer. The Algebra I class brought in a toy car and boat and used my desk as the cliff. They used a couple shoe laces as the rope, moved the car forward and measured. The Geometry class used the Pythagorean Theorem and the Calculus class used an equation in which they then took a derivative. Students in all three classes came up with accurate solutions by three very different methods and were able to convince their fellow classmates that their solutions were correct. Perhaps the reader can come up with another method which yields a correct solution.

The Pythagorean Theorem is one of the oldest and most important theorems in geometry. Most high school students state the Pythagorean Theorem as $a^2 + b^2 = c^2$. However, the Pythagoreans did not have the advantages of our algebraic representations. They thought strictly of its geometric representation. In a right triangle the square on the hypotenuse is equal to the sum of the squares on the other two sides. I asked my students to draw equilateral triangles on each side of a right triangle. Is the area of the triangle drawn on the hypotenuse equal to the sum of the areas of the triangles drawn on the other two sides? After they had proved it is I asked them to draw semi-circles, then rectangles twice as long as wide, and finally hexagons on each side. Would the theorem still be true? The students quickly proved that the theorem would be true as long as you had similar figures on the sides. Thus they had a more general theorem than the Pythagorean Theorem. Unfortunately there is no easy way to write this general theorem algebraically.

When teaching the Pythagorean Theorem, I always liked to explain how the American Indians discovered the theorem for themselves. Three pregnant squaws entered an Indian hospital at the same time. One squaw was given a horse hide to sleep on, the second was given a bear hide to sleep on, and the third was given a hippopotamus hide to sleep on. The squaw on the horse hide and the squaw on the bear hide each gave birth to a beautiful son. The squaw on the hippopotamus hide produced twins. Thus, the squaw of the hippopotamus hide was equal to the sum of the squaws on the other two hides.

The Indians were also famous for the mnemonic device that many students use for memorizing the definition of the first three trigonometric functions. (SOHCAHTOA) A small Indian boy went on a hunting trip with his father and uncles. He happened to trip and hurt his toe, and thus he limped around all day until he returned home. Then he went to the hospital tepee. There the Indian doctor instructed him to get a large tub and fill it with hot water and then sohcahtoa.

On the last class before Christmas vacation I always had a mathematical Christmas card on the board as the kids entered the room. It was up to the students to decipher what the card said. It simply read

ABCDE
FGHIJ
KMNOP
QRSTU
VWXYZ

On the test at the end of a section on trigonometry I always had bonus questions. I always included my favorites.

Which trigonometric function would best describe each of the following?

1. A highway billboard

2. A gentleman at the beach

3. The second person to sign a legal document

4. A blind man

A few more questions of this type I used were

5. What Roman numeral can climb a wall?

6. How can you take half of twelve and get seven?

7. What one English word does almost every American pronounce wrong?

For Algebra I students who had trouble factoring I stated there are only five methods--common factor, difference of two squares, trinomial, sum of two cubes and difference of two cubes. (For math teachers reading this book, I claim grouping is a special use of the above five methods.) I further insisted that the easiest way for students to factor an expression is follow the order in which they learned the five factoring methods. When a student missed seeing a common factor in class I always exclaimed, "When factoring, the first thing you ALWAYS look for, no exception, is a common factor!" After I had said this refrain a number of times, when a student missed seeing the common factor, the entire class would chant, "In factoring, the first thing you ALWAYS look for, no exception, is a common factor."

Students often used FOIL (first, outside, inside, last) to remember how to find the product of two binomials. I asked whether that method would always work. Just because I had asked, some students thought there had to be an exception. There isn't, it always works. I love rules that always work, no exceptions. That is, FOIL always works provided you are multiplying two binomials.

Then I asked this series of questions. What would the answer be to $(X)(-X)$? Everyone knew the answer is $-X^2$. Do $X = X - 0$ and $-X = 0 - X$? Of course. Then we checked using FOIL. Usually the inside and outside terms combine so we included an addition sign there.

50

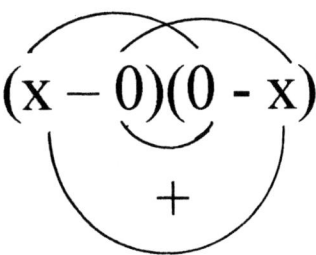

$$(x - 0)(0 - x)$$
$$+$$

Most of the time I would add a pipe and top it off with a hat.

When the class began to study the quadratic formula, I went over the derivation and assigned many homework problems to be solved using the quadratic formula. The next day as the students appeared at my door, I told them they had to know the magic word to get into the classroom. After a few guesses like "abracadabra," they realized what they must say. If they missed it, then they had to step outside and study until they could recite it correctly. Of course hearing their classmates spout it off helped them to get it right. On the next test involving the quadratic formula, every other question was, Write the quadratic formula. Those students who did not know the formula failed the test. I have on occasion told my classes that the first question on the final exam would be--Write the quadratic formula. If a student got it wrong I would not look at the rest of his or her examination. I do not know what I would have done if someone had ever missed that problem.

Most summers I receive at least one postcard from a former Algebra I student with nothing on it except

$$x = \frac{-b \pm \sqrt{b^2 - 4ac}}{2a}$$

Most years, over a long weekend, I would give an interesting problem that the students could enjoy. One of my favorites was the gas-water-electric problem. On one side of a piece of paper I drew a gas company, a water company and an electric company, all in one row. On another row I drew three houses.

Gas	Water	Electric
House 1	House 2	House 3

The problem was to draw a line from each company to each house. No lines could intersect and a line could not go through a company or a house to get to a destination. The solution had to have nine lines all drawn on one side of the piece of paper.

The next class period the kids always had to ask how to solve it. I must admit that I was amazed at the cleverness of some of the students who attempted the problem. Every year I had one or two students take the paper, put it on a globe, and then try to draw the lines. Although this method did not work it was a step in the right direction.

If there were a few minutes left in a class, I always liked to tell some little math story. In the beginning of the year students thought when I started out that the story was true, but they soon learned that it probably ended in some kind of joke.

Early each year I wanted to know if any of the students were taking French. I always had a few. So I told them a sad tale. You see, my grandmother lived in France and had a great love for cats. At the time she was going to make a visit to the United States she owned three cats. She was not imaginative with names so she called them un, deux, and trois. She hired a cat sitter to care for her cats while she was on her trip. It so happened that behind her house was a lake and while she was away, the lake froze over with a thin layer of ice. The three little kittens wandered out on the ice and fell through. When my grandmother returned, the cat-sitter struggled to explain what had happened. Finally, she said "The cats walked out on the ice, the ice broke, and then un, deux, trois, quatre, cinq."

Sometimes, when I had about ten minutes left at the end of a class period, I suggested we play Who Wants To Get Out Of Class Early?, a takeoff of the television show *Who Wants To Be A Millionaire?* My rules were that a contestant answering three consecutive multiple-choice questions correctly could be excused from class or attempt another set of three questions. If the contestant answered all six questions correctly then he or she could take a friend with him or her. However, if the contestant answered three additional questions correctly, the entire class would be excused. No one ever answered all nine questions correctly. Of course I made the last few questions so difficult that a contestant could only guess the correct answer. He or she only had one chance in four of picking the right answer. Most of the students used up their lifelines early in the game. It was interesting sometimes to see whom a student would choose for the call-a-friend lifeline. On more than one occasion I am sure that the "friend" knew the correct answer but gave the wrong choice just so the contestant could not get out of class early. It was always done in good humor.

When teaching the laws of exponents I always wanted a student to simplify the following expression.

$$\frac{I^{10}C^{12}E^{14}}{I^{7}C^{9}E^{11}}$$

The answer was to be written in the form of $(abc)^n$.

When my Algebra II classes studied logarithms, I always put the following extra credit problem on tests.

Simplify into the form of one logarithm.

$\text{Log}(C) + \log(A) + \log(B) + \log(I) + \log(N)$

I tried to liven up the study of calculus with such problems as

$$\int_{0}^{ice} 3x^2 \, dx$$

$$\int_{1}^{cabin} (1/x) \, dx$$

$$\int (1/cabin) \, d(cabin)$$

At the end of a calculus course, I always put a problem on the chalkboard which the students were to read out loud. I explained that this was a test to see who was a real math student. Surely a true math student could read the following problem correctly.

54

(Note: I purposely left off the dx in the integral. Many students do so also, but not on purpose.) I did always review the integral sign, the exponential function, and the functional notation beforehand.

Then I wrote:

$$\int e^x = f(u)^n$$

After the entire class had answered "The integral of e to the x equals a function of u to the n," I checked out in the hall to bring in the youngest student I could find. Since he knew no calculus, he immediately read "Sex equals fun."

For several years I taught a geometry course without using a textbook. The students wrote all their own definitions and derived their own theorems. We discussed possible theorems in class and for an assignment they would try to prove them. Of course, sometimes they had what they thought would be a theorem and it was not true and of course it was impossible to prove it. I believe the Foreword in *Geometry According To Us* best describes the course.

Several years ago a colleague and I were discussing how to improve the geometry program taught at The Hill School. We came back to this discussion on several occasions and this book is a result of what we thought would be a great geometry course. My colleague, Dennis Buzzard, passed away last year. I am sure this program would have been vastly better, if he were here to give me the additional help and advice at the difficult times during the school year.

The class was hand picked as an honors section of first year geometry students, three sophomores and six freshmen. The book is not meant to be a text, but rather a compilation of what these nine students learned during the past year. There was no textbook. *The Thirteen Books of The Elements*, translated by Sir Thomas L. Heath, was used as a reference guide, sometimes used as a crutch more than I would have preferred. In many ways the geometry of these students is similar to the current geometries taught to high school students throughout the nation; yet quite different. The class was allowed to make their own definitions and assumptions as long as they were consistent. Naturally, throughout the year these geometricians would have to go back, revise their definitions and reword axioms, postulates and theorems. Superposition was accepted and thus all the triangle congruence theorems were proved. Many readers may question my allowing superposition to stand. I never said no nor gave an explanation, but my questioning probably was often leading when they wandered too far off track. Their decision was always final. Thus superposition stands. You will not see the reflexive property of equality, but instead the superposition statement "Things which coincide are equal." Superposition cannot be that bad as Euclid himself accepted it in the use of a proof.

Every teacher looking through this book should know that I have never had more fun in teaching a course. It was much more work than I had originally thought, but it was worth it. I certainly plan to do the course again. The students permitted me to write this Foreword; from this point on, however, everything is theirs. I take no credit and accept no criticism for their work. There are mistakes, but I know their work is vastly superior to anything I could have done at their age.

This book could not have been possible without the help of many people. To list everyone by name would take too much space, but several need to know of my appreciation for all they have done. To Mrs. Sara B. Harris, an extraordinary teacher, my gratitude for enlightening me in the beauty of the mathematical world. To Sidney E. Ainsworth, my first and greatest department chairman, for instilling in me a love and care for teaching. To Judy Borger, Larry Kelly, George Kovac, and Jim Long, our appreciation for helpful suggestions. To the Mathematics Education Trust for their financial support: without their help this book would still be my dream. My greatest thanks go to the people who did most of the work, made the experience meaningful, and made my dream a reality. Therefore, my heartfelt thanks to Tony, Damon, Joe, Patrick, Frank, Phil, Robert, Rooshabh, and Janak. Without question, you are the most memorable class I ever had. To the reader – Enjoy *Geometry According To Us*; I did!

One theorem in that book which does not appear in most high school geometry texts is "If two triangles have a corresponding angle and two corresponding sides equal, then the triangles are congruent, or the triangles have a pair of corresponding angles supplementary." I received a phone call one night from a math teacher who had come across the book and wanted to congratulate me on the project. He also wanted to know who the mathematician Fitzgerald was because he had researched many math books and could not find that name. I explained to him that the mathematician who proved that theorem was Joe Fitzgerald, a student in the class. Thus, it was named the Fitzgerald Theorem.

At the end of each school year I invited each of my classes to my house for a hallfeed. I always served cake and a punch of orange sherbet and ginger ale. This hallfeed always included a tour of my house so that I could show the students my baseball and autograph collections. They always seemed to enjoy listening to my stories about unique items. I have an old Hill School baseball uniform which in the early 90s was found in a foot locker in the gymnasium. There were actually two old shirts in this locker. During Strawberry Festival at the end of the school year, the Athletic Department sells the old equipment it no longer needs. I bought both shirts. Howie Bedell, a former major league outfielder with the Braves and Phillies, known mostly for breaking up Don Drysdale's string of scoreless innings, heard that I had bought two old uniforms. He asked if I would trade one of them. I agreed. He said he could give me a baseball which he had The Hill School baseball team sign. I said fine. He looked at me as if I were crazy and asked if I meant it. I said yes because he had given me so many baseball souvenirs already that it was a fair deal. So he agreed to bring the ball down the next night. If ever there was a class act, it is Howie Bedell. He arrived with two shopping bags full of baseball memorabilia for me. There were score cards, magazines, a Cincinnati Reds rain jacket, some bubble gum cards, and of course the baseball signed by the Hill School baseball players. I claim the two uniforms I bought were worn by Ty Cobb when he helped coach The Hill School baseball team in the early 1930s. I do have a photo from *The Hill News* of Ty Cobb wearing a uniform that matches the one I kept. Why else would Mike Sweeney, Athletic Director, have stored away these two old uniforms in a locked trunk? It took more than fifty years for that trunk to be found and opened, but the uniform is now proudly on display in my "dugout."

I also have a six foot bat. Every Christmas at The Hill School Christmas Dinner, one of the students dresses up as Santa Claus and gives out joke gifts. Usually I received a comb or a pack of baseball cards. However, one year the students made this bat in the wood shop and presented it as a joke about my baseball collecting. I think they were disappointed that I thought it was great and displayed it for years in my living room. Today it is displayed in my computer room along with my bobbin head doll and Hartland Statue collections.

Whenever I gave students a tour of my house, I always pointed out autographed baseballs, photographs, pennants and even items from the old Pottstown Firebirds, semi-pro football team. My book collection includes a copy of the book about the Firebirds, titled *The Forgettables*. I always had to read the first sentence in this book out loud to the group as it is my favorite line and so true. "You would never arrive in Pottstown, Pennsylvania, by mistake." It then describes the terrible road getting to Pottstown from the Pennsylvania Turnpike.

Once a student went home and described my house and autograph collection to her father. He stopped me at his daughter's graduation and said that when he had been a student at The Hill School, one of the teachers had given him a letter from Herbert Hoover. He asked if I would like to have it. Obviously I would, and he sent it to me a few days later. I had it framed and it now hangs in my hallway.

A student once remarked that when kids came to my house for a hallfeed, it was like another class, just not mathematics. They learned so much about so many different things. Little did they know, I probably learned just as much from them.

I am convinced that sharing my home and hobbies with the students made for closer relationships and in the end a more fruitful learning environment. The use of many tricks, puns, and games used in class helped to make learning an enjoyable experience for the students. When students are happy and enjoy what they are doing, they do their best work. The students seemed to appreciate what was done for them and I really enjoyed sharing a total learning experience with all my students.

Chapter VI

The DLSF

In 1984-85 I taught two sections of Algebra I. These sections comprised third formers and just three second formers, Chris Drowne, Tim Robertson, and John Marcus. The Hill accepted approximately twenty students each year into the second form. Usually these young men were not only gifted but also quite mature. Chris, Tim, and John were arguably the best students in the class. Unfortunately for them, I happened to choose all three of them as the students I would pick on that year. They would be called on more than anyone else and that it would be very obvious if ever they were not fully prepared. Whenever there was a question I was almost sure nobody could get right, I called on one of those three students. Again, I think this attention was a great benefit to them as they soon learned to think outside of the box and to stay one step ahead of me. There were occasions when all three came up with correct answers to very difficult questions.

One day all three of those second formers had a particularly bad day. Nothing seemed to go right. I commented that these three students were just dumb little second formers. Others picked up on the comment and whenever Chris, Tim, or John missed a question, someone would blurt out that he was just a dumb little second former. The expression was used so much that eventually it was shortened to just DLSF.

The first thought I am sure the reader has is that that is an awful expression to use with school kids. On the surface it may be. However, because of my relationship with these three students, everyone knew that I used it in fun. In fact, the expression has become one more of endearment than of ridicule.

Since then every second form student has lived with the label DLSF. I mean *every* second former, even if he had not been assigned to my class. Because I usually taught the Algebra I Honors course I usually had a handful of DLSFs in class. Evidently word got out, even to the new students, and spread around the campus. I remember one day in the first week of classes a second former made a mistake and I said you must be a DLSF. Of course, he asked what a DLSF was. Another second form student, Matt Adukaitis, explained it to him. He even said that I would probably be making fun of them all year. Matt did not realize that I would use that expression their entire Hill School careers. I often said once *a DLSF, always a DLSF.* Therefore every student entering The Hill School in his second form year is forever a DLSF. As the years go by former second formers have become more and more proud of their labels as DLSFs.

On Parents' Weekend, all parents are invited to attend classes with their sons and daughters. I had to smile as Matt Adukaitis' father walked into my classroom. He had turned his name tag around and had rewritten simply "DLSF PARENT." All day he proudly wore this name tag on his chest.

Sometimes when one did something wrong he would explain to me that I could not expect otherwise because he was only a DLSF. I received many letters from former students during their summer vacations and long after their graduations. Many of these letters had the closing "your DLSF" or "your favorite DLSF."

The Hill School no longer has DLSFs. The School discontinued the second form when it began to admit young women. Therefore, the last DLSF graduated in June 2002. It was quite appropriate that I chose that same June to retire. What would I do at The Hill, if there were no DLSFs to pick on?

For those DLSFs I had in class their first year at The Hill School, I would try to collect and save everything in which their names appeared: *The Hill News*, the *Pottstown Mercury*, sports or theatre programs, or even notices posted around campus. I also tended to take more photographs of them than of other students on campus. I had reason as I could use these photographs when I put together scrapbooks. Early in my teaching career I had difficulty coming up with nice, inexpensive graduation gifts for those students I had become very close to. I could not afford to spend $25 per student to buy gifts. A scrapbook seemed ideal. The books would cost me about $5 apiece, and I could fill them with memorabilia from throughout the students' careers. I would make copies of their report cards and save any notes they had written to me. After the introduction of e-mail on campus, I had many, many notes to add to the scrapbooks.

I never said anything to the students about making these books. I believe a few had their suspicions because they had heard rumors from older students who had received scrapbooks at their graduations. But, not many knew. I always gave out the books on graduation day, and most of the younger students in the School were not around the sixth formers on that day.

On opening day and several other occasions during the school year all students are given name tags to wear. When I asked a student for his discarded name tag he probably did not know why I wanted it. He would ask, but I would usually say something about I am not positive right now. I would assure the DLSF that once I had decided what to do with the possession he had given me that I would tell him.

Sports Awards assemblies are held at the end of each term. At this ceremony each coach talks about the team and then presents letters and special awards. One year I thought it would be interesting to see whether any students would give me their sports letters as they exited the ceremonies. After the Fall Sports Awards ceremony I stopped fourteen DLSFs, some I had in class at the time and others I had in previous years, on their way to their next classes and asked what they planned to do with their letters. Then I added, "If you are only going to throw it in a drawer and forget about it, how about giving it to me." I was dumbfounded when thirteen of them actually gave me their letters on the spot. Chris Clarke, a DLSF I then had in class, said that this was the first letter he had ever won so he wanted to keep it. However, if he won a letter in hockey in the winter, he would give that one to me. The next day, when he came to class, he had the letter with him. He handed it to me saying that if I really wanted it, I could have it. He gave me several other letters during his career at The Hill. Sam Carella, who was in the same class as Chris, had an outstanding career as a runner. He gave me every one of the letters he had won in cross country, winter track, and spring track. When he won his first medal, I had asked him if he would give it to me. I was awestruck when he handed it over. During his career at The Hill he gave me 14 such medals which he had won in his track endeavors. I put them together in a display box and gave it to him at his graduation, along with his scrapbook, which included all of his sports letters.

In some ways I was a little disappointed that students would give up their hard earned letters so easily. Later Rick Hilton, Dean of Academics, made me feel very good about it. He claimed that he could stand outside that auditorium and ask for letters and probably would not be given a single one. He thought those DLSFs were giving them to me because I wanted them, not because they did not want them.

I also asked students for their acceptance letters from colleges. I asked for copies if they did not want to give me the originals. An acceptance letter makes quite a display in the scrapbook and is something each DLSF will have for the rest of his life. I wonder how many readers still have acceptance letters from colleges.

When I first began the scrapbook project, I used only the student's last year. It was difficult to collect enough material to fill the books. I then changed to collecting material beginning with the DLSF's first day on campus. Once I became interested in photography and began to take lots of photos, I had no problem filling a book. In fact, last year, May 2002, a number of students had two and three volumes to lug home. The downside was that there were a number of DLSFs I did not really know well until their second or third year at The Hill. It was too late then to begin collecting for a book. I always felt bad when that happened.

Most of the DLSFs usually wrote nice thank you notes after graduation, although I do think the parents appreciated the scrapbooks more than the kids. However, I am sure a number of years later each student will appreciate his book more and more. Here is a sample thank you I received this past year. "I wanted to thank you for the album that you assembled. That gift means more to me than any other I received. I don't know how you did it, but you managed to put so many memories into one package. I will miss hearing your DLSF jokes around campus. I also want to wish you good luck on your retirement, although I can't really picture The Hill without you. I hope we can keep in touch, who knows; maybe we'll meet in Cooperstown one day. Thank you again."

Chapter VII

Saturday Night Survey

Six distinguished DLSFs graduated in May 1995. These six students became known as the DLSF 6. I had all six of them for an Algebra I class in their second form year. In the ensuing years I had these six students a total of seventeen times for various math classes. In addition, they were regulars for extra help and hallfeeds. They were a special group, and I still stay in close contact with most of them.

In the fall of 1995 I wondered how many college freshmen would answer e-mail before midnight on a Saturday night. I sent an e-mail to each of the DLSF 6, stating that I was doing a poll. I did get two answers before midnight. I received two more answers around 3:00 a.m. These two stated that they had been out and had checked their e-mail before going to bed. The other two responded during the day on Sunday. On subsequent Saturday evenings, I sent similar e-mails to the DLSF 6. Then I began to add the latest Hill School news to the survey. During the winter I received e-mails from other graduates of that class of 1995, asking that they be put on my Saturday Night list. These graduates had talked with several of the DLSF 6 and were astonished by how much the DLSFs knew about what was going on at The Hill School. So I included these additional graduates to my *Saturday Night Survey* subscribers list.

I thought there may be other college students who would be interested in my *Saturday Night Survey, (SNS)*. I e-mailed those Hill graduates I knew quite well to find out if they wanted to be added to the *SNS* list. Then I began to ask a question on each *SNS*, one that had no right or wrong answer. Usually those early questions were ones of moral judgments or just routine things like what was your favorite Hill School meal? This whole idea then mushroomed as I added former faculty and some parents to the *Survey*. When I commented in my classes about some of the results, many current students asked if they could be added to the list. I added anyone who wanted to join. Many of the original group have now graduated from college, but have asked to remain on *SNS*. Today, members include current students, graduates, faculty, former faculty, girl friends, and friends of the faculty. It is a good way to keep in touch. *Saturday Night Survey* has one hundred and sixty subscribers. The average number responses per week are one hundred and ten.

I did add a rule. If a member went three consecutive weeks without responding, I remove his or her name from the list. I stuck to this policy, but whenever anyone wrote to ask if he or she could be returned to the list, I always agreed to do so. Still. most of the *Survey* members are college students and *SNS* remains primarily for them. Today a question goes out every Saturday and those who answer by Tuesday midnight get to see everyone else's answer and receive the latest Hill School news..

I call the Wednesday edition *Survey Says* and send a copy to those who responded to that week's *SNS*. Respondents' names and answers are listed in the order they were received. The last section is the "PGNUS" section. When I talk with some of the members of *SNS* they always pronounce it PIG-NUS. Originally I did explain the section to everyone: The "*P*" is silent and "GNUS" is pronounced like the animal. Therefore this is my news section.

I have also added recently an *SNS News Flash*. I send it when an urgent news story simply cannot wait for the following Wednesday's PGNUS. Very important announcements I send out as a *News Flash* because not everyone responds every Saturday night and I want to be sure everyone receives the news. *News Flash* events can be the birth of a faculty child or the death of a Hill family member. Death notices particularly must get out immediately as many of the people want to attend the services for the deceased. Many college students who cannot get away want to write to the family, so I try to include a name and address in my *News Flash*.

A favorite *SNS* question actually had two answers I will always remember. You stop in a camera shop about 50 miles from your home. You did not go there for the purpose of visiting that shop; you just stopped in. You look at the camera bags because you need one. The clerk comes over and asks if he can help you. You state that you like a particular bag but you are not sure your equipment will fit. The clerk asks what camera equipment you have and then proceeds to pull off the shelf items similar to yours and shows you how they would fit into the bag. You buy the bag and when you get home you realize the sales clerk has left a $300 lens in the bag. What do you do? Brian Craighill, one of the DLSF 6, who probably knows me better than any other student I have had--and even better than I had realized--wrote the following: "By the way you asked this question, I think this probably happened to you. I know what you would do. I hope I would do the same." He was correct; it did happen to me. When I took the lens back, the salesman and the owner of the store were there. When I explained what had happened and gave the owner the lens, he commented, "I do not believe this; you are one in ten thousand who would bring this back. We would not have discovered it missing until next spring, when we did inventory, and would have no idea what had happened." He thanked me for doing the right thing. For this camera question one imaginative young man wrote that he would go back and buy another bag.

Often I got students who gave humorous answers. Some of the unique responses came from individuals who never attended The Hill School as students. Amy Pentz is the wife of a former Hill graduate. Kate Carella is the sister of two Hill graduates, but she did not attend The Hill. Julie Horvath is the daughter of a former Hill School faculty member. I look forward each week to read the answers from these three people.

Following are some of the questions I have used and some of the responses that I thought for some reason were unique or interesting.

Question: There will be a tribute for Mr. Thomas Ruth at his retirement on Alumni Weekend. What is your favorite story or anecdote about the TRuth, Mr. Ruth? (You may submit more than one answer.)

Roy Woo wrote, "I always used to complain that Mr. Ruth never gave the typical pizza hall feeds. Instead, he treated us to a fine selection of teas and pastries almost every Sunday. What I wouldn't give to have that every Sunday now!" Fred Bobb responded, "I asked him: 'Mr. Ruth, why did you become a master at The Hill?' I will never forget his response. He said, 'for three reasons, June July, and August.'" Janak Vidyarthi wrote, "I was a third former on Mr. Ruth's hall. The School closed the building one afternoon because they were going to search everyone's room because a faculty wife had had some clothing stolen. Mr. Ruth was opening all the drawers in my room and closing them without even moving anything or looking under my clothes. So I said, 'Mr. Ruth, if I was smart, I would hide her clothes under my clothes. You're not even looking.' Without missing a beat Mr. Ruth replied in all seriousness: 'I don't think you're smart.'" Bill Barnshaw replied, "During my fourth form year I traveled to London, England for the first time. One day while walking through the British Museum, I see this gentleman walking slowly through the lobby. I shout 'Mr. Ruth' and slowly but surely he turns around and mumbles a profanity to himself. Later that week we met up for dinner at the infamous Sloan Club. Moral of the story, Mr. Ruth can't go anywhere without being accosted by Hillies." One of my all-time favorites was Doug Brody's reply. "I was visiting The Hill in its last year before coeducation. I applied late, and my visit was not during the regular visiting time, so I was the only visiting kid on campus. After study hall, I was tossing a lacrosse ball around with a student in the dorm and I guess we were being a little too loud. All of a sudden, I heard a grumbling sound of discontentment from the faculty apartment, and it was then that I saw the TRuth for the first time, enraged and pointing a stick at us. He said something to a prefect, who informed us all of a mandatory meeting in the commons room. We all went in, sheepishly hanging our heads, and were told to sit down. Mr. Ruth informed us that the involved lacrosse equipment would be impounded, and that furthermore we were all to receive demerits. To receive our demerits a sheet was passed around, which we were all to sign. When the sheet got to me, I looked up to inform Mr. Ruth that I did not attend The Hill quite yet. He gave me a cold stare and said, 'Sign the sheet; you'll get the demerits when you get here in the fall if you don't get them now.' And thus my clean disciplinary record was spoiled before I arrived."

Question: If you could communicate with any type of animal, which would you pick?

Mr. Warner, a former math teacher at The Hill, responded, "One that was about to eat me." Scott Detar wrote, "Dogs – because they watch us do the stupidest things ...I would love to know what they think of us." David Zuckerman, a DLSF, wrote, "Right now I'm just trying to communicate with humans, so the thought of communication with other species is a bit out of my reach for now." Dan Sutko replied, "Cats... just so I can ask Socks all sorts of fun things." Kate Carella, one of the respondents who always has a unique reply, answered, "Definitely my stuffed animal. I think I had the capability when I was younger but it was somehow lost along the way. It would be nice to have it back." Mr. Kelly, a great math teacher, responded, "My students." Amy Pentz, a Hill graduate's wife, answered with one of her usual witty remarks, "The infant human being so I could know what the hell was wrong at 2:00 in the morning." Jamie Knise, a DLSF, replied, "My dog because I would want to know why it is such a retard."

Question: If you could revise the income tax system, what would you propose?

Lawrence Clark, obviously a DLSF, wrote, "No taxes on DLSFs." Mr. Kovac, a member of the Mathematics Department at The Hill School for too short of a time, replied, "Any tax increase would be voted on in Congress and if your congressman voted for it, people in that district would pay the increase. Those whose congressman voted against it would not have to pay. This would serve to keep taxes down or change congressmen quickly."

Question: Describe your most rewarding experience!

Erin Romig, one of the first girls to enter The Hill and in my opinion an honorary DLSF and a master of sarcasm, replied "Joining *SNS*; it brings a ray of joy to each Saturday of my life!" Brendan McGowan, a current student, knew that if he did not reply he would not receive a copy of *Survey Says* the following Wednesday. So he simply stated, "I don't know, can I just see the answers?" Amish Patel, a true tormentor of Mr. Ruth, always seemed to have a reply that involved his favorite teacher. His response was, "By far the most rewarding experience was living on Mr. Ruth's hall."

Question: What is the greatest or most interesting experience (good or bad) that you have had with snow?

Philip Gallagher, who never became a Hill student, but did attend the summer program and always seemed to have a smart aleck remark, wrote, "I lost a model car in the snow one winter in December. I found it largely intact in March once the snow finally melted." Sam Carella, a DLSF, probably saw more of me than any other student at The Hill. I had him in class twice, and was his adviser for five years. He also attended the summer program at The Hill twice. During his first summer at The Hill, I bought my camera. The first photo I took was of Sam and his sister walking across the campus. Even though Sam's sister Kate never attended The Hill School except in the summer, she usually responded to *SNS*. It was quite interesting to see the similarities in the answers of these two very close siblings. Sam's response to this question was, "'Getting bombed,' as the caption under the picture in the newspaper called it, by my sister was most definitely a bad experience." Kate Carella replied, "One winter my brother and I were playing in the big snow storm and we had a snow ball fight which we each swore we had won, but there had been a newspaper photographer and reporter going through the town to get people's reactions to the snow storm. The very next day there was a huge front page picture of me hitting my brother with a snow ball…. Underneath, the caption read 'Sister bombs brother with a snowball.' We all know the winner of that snow ball fight now, don't we?" William Bouvel wrote, "Hiding my ring in it last February. No one ever found out." (I still do not understand why this DLSF would be throwing snowballs without gloves while wearing a loose fitting ring.)

Question: What specific aspects of being a child do you miss the most?

Philip Gallagher often wrote answers knowing that his mother was also a member of *SNS* and would read his responses. He wrote, "The chauffer. (And no one is expecting you to be able to spell words properly.)" Damien Newton responded, "Having other people make dinner." Mr. Yoo, a science teacher and graduate of The Hill wrote, "Playing outside without any worries. Where did all the stress come from?" Tom Power, former Hill School class president his sixth form year, now as bald as I am, wrote, "A full head of hair" Mr. Ruth, better known to all SNSers simply as TRuth, wrote, "I don't miss anything. Unlike some others, of a very old nature, I am close enough to that time that I really still have fun. I can't really say that I miss it because I still do it. I loved nap time as a kid, and I still love it."

Question: What specific subject do you feel that you know better than any other subject?

David Zuckerman, obviously one of my former math students, replied, "I'd like to say math, but I'm sure you'll argue that." X Prines, a student I did not know well while he was at The Hill, wrote to ask to be included in *SNS*. He always provided us with an interesting comment. His reply here was, "It's a toss up, TV, eating, or looking at beautiful women." Tom Power provided us with another of his witty remarks with, "Humility. Although I'm still not very good at it."

Question: What is the most memorable phone call you have ever received?

Ryan Yerger replied, "Hello, there is a 99.97% chance you are not the father." Courtney Steltz, a very funny person and always with a smile, provided SNSers with more serious and thoughtful responses. Her answer here was, "My uncle had struggled with cancer for two years and the doctor had given him hours to live. Because he was in Florida and I could not be there with him, my aunt called and she held the phone to his ear as I told him goodbye and how much I loved him." John Hurley wrote, "Santa Claus called me when I was a young one." Mr. Borger, a faculty member who arrived at The Hill the same year I did, answered, "It came from our son in his sophomore year of college. He said very simply, 'I think I have met the girl I want to marry.'" (A few years after this phone call, I attended the wedding of John and Molly Borger.) Mr. Kelly will never let me live this incident down. When I wrote the question, I knew what his answer would be. "It came when Mr. Pierre offered me a job at The Hill, I had told him about the 6 hour time change, and so if he called around 4 p.m., I would be home because it would be 10:00 p.m. in Belgium. So he called at 10:00 p.m. Pottstown time."

Question: Ground has been broken for the new athletic center. Your task this week is to name the new building.

Mr. Reifsnyder, a history teacher at The Hill, wrote, "The Mercer Athletic Center; it will be The Hill's own 'Big Mac.'" The building was named the David H. Mercer Athletic Center in honor of The Hill's long time Athletic Director. Over 60% of the *SNS*ers chose him, rightfully so, as the person for whom the building should be named. Mr. Mercer is a grand-fatherly figure not only to all the students, but to all on the faculty as well.

Question: As a child, what was your favorite pet?

Jamie Knise, a DLSF, wrote, "My favorite pets were my fish, but they all died because my cat ate them." Again, Kate Carella had one of my favorite replies: "My favorite pet as a child was DC -- my kitten! She was named DC because my mom hates cats (Don't ask how we convinced her to get one.) and so she got to name it.... 'Damn Cat' = DC. That was a funny story, taking DC to get blessed at the church as the priest inquired, 'Oh, did you get her in Washington, DC?'"

Question: What was the most significant thing you learned this past summer?

Adam Weller, another DLSF, responded, "How to work in the real world." Erin Romig, who received my grief continuously as an honorary DLSF, replied, "Finish summer reading BEFORE summer so I don't have to put up with Mr. Pierre's emails nagging me every day!" Philip Gallagher replied, "Lead paint: Delicious, but deadly." Mr. Warner answered, "I'm a whole lot older than I hoped I was." Mrs. Epps, mother of Jamie Knise, answered, "The most significant thing I learned this summer was the reinforcement that all in all I have raised a wonderful son. (With the help of The Hill of course.) It was very hard to send him off to college even though he had been 'away' from home (20 minutes), but I know he has learned his lessons well and he will do marvelously! Thank you for your help folks.... And you know who you are!" Mr. Kovac replied, "Never buy chocolate as a gift and go to Palm Springs, CA, in the summer before you give away the chocolate." Jim Hollister, one of the DLSF 6, responded, "That the customer isn't always right." Julie Horvath, daughter of Mr. and Mrs. Kelly, often gave interesting responses. She phrased her answers to get her parents' attention, knowing both of them were members of *SNS*. Her reply, "Patience at 3:00 a.m. when your baby is screaming and she has eaten, been changed, and still isn't happy or sleeping." Mr. Drowne (the real one) graduated from The Hill in 1964. His oldest son is a Hill graduate and current faculty member teaching history. Because most students would think the son, Mr. Chris Drowne, was giving the answers, I have always identified his father, Mr. Rusty Drowne as 'the real one.' I have probably gotten to know this family better than the families of any of my other students. The real Mr. Drowne's response, "I learned first hand not to believe the hype about Arizona's 'dry heat.' When it's 110 degrees, its 110 degrees... period. Enough of this 'dry heat' nonsense." Keith Krem, a DLSF, wrote, "I learned not to throw parties at my house when my parents are in Europe." Mrs. McCallum, mother of two

Hill graduates, replied, "I am too old to baby-sit a 3 year old and an 11 month old for 3 days." Dan Confer responded, "I learned the real reason I am in college. I worked in a lumber yard. `nuff` said." Ryan Yerger wrote, "No matter how hard you try, there are some things you simply can't do." Paul Weaver, a DLSF who grew up in Pottstown, replied "There is really nothing to do in Pottstown, Pennsylvania." Kevin Hyde, a DLSF, wrote, "I learned that lying to employers is not the greatest idea." X Prines used this forum to solicit a job: "Having only a 3.0 makes it difficult to find a fun mechanical engineering job. (Hint to anyone looking for someone graduating in December.)" Mr. Dollhopf, a Hill math teacher, replied, "There is nobody more important in your life than your family." Mr. Dollhopf's wife is also a math teacher at The Hill and a member of *SNS*. Their responses often revolved around family. They are two great teachers who truly make The Hill School what the founding father wanted, a family boarding school. Andy Basco, a DLSF, obviously played some cards during the summer and answered, "A full house beats three pair."

Question: What food would you rank first on your list of least favorite foods?

DLSF extraordinaire Mike Weigley answered, "I would have to say anything green." Chris Hagge replied, "Hill School Shepherd's Pie. A meal for which it was definitely worth saving one's demerits and sneaking to the best restaurant in Pottstown, Little Italy. (Or so I have heard, for I would never have left campus during school hours.)" Matt Holt responded, "Anything that you have to wait in line for."

Question: Help me do my Christmas shopping. Please give one good suggestion of a great Christmas gift.

X Prines answered, "I don't have an answer, but please be sure to send the answers ASAP. I also need help." Alex Blood responded, "A shirt that says my two favorite teams are the Redskins and anyone playing the Giants." Doug Bouquard, who was in my Algebra II class at the time, wrote, "The answers to tomorrow's test."

Question: Do you know where the sports letters or academic awards you won in high school are today?

JC Groon responded, "Sure, they're all in my mother's house safe and sound ... Somewhere." Shawn Fernandes, an outstanding track star, replied, "I would still wear my letter sweater today, but because I won it in third form it doesn't fit anymore." Mr. Drowne (the real one) replied. "My 36-year-old letter sweater is in the attic. It's in good shape, except that it has shrunk a great deal so I can't wear it." Mrs. McCallum answered, "What letters? What awards?" Charlie Evans, a student I had obviously asked to give me his athletic letter, answered, "If you really want my j.v. water polo letter you may..." Mrs. Dougherty, wife of the headmaster, replied, "Like most mothers, my mom threw away some of my things." Tim Kelly was an outstanding photographer but certainly not a jock. He simply replied, "Never got one." Philip Gallagher, again holding his mom responsible, answered, "I imagine mom has those someplace, but I am not too sure. I'll ask her."

Question: A) Where did you have your Thanksgiving dinner?
            B) If you did not have turkey, what was the main course?

Peter Shea replied with this gratifying answer. I was pleased to see that he did indeed celebrate the holiday even though he was so far away from the United States. He wrote, "Believe it or not I did have a turkey Thanksgiving dinner here in St. Andrews. My flat mate puts together a dinner every year for American students at the University. It took place at the Scores Hotel. A note of interest, when he first approached Scores Hotel management in 1994 they had no idea what Thanksgiving was and he had to explain what it is and what one eats for the holiday." I always made fun of Lisa Huyett for being from Oley, a village which probably has more cows than people. Her answer was, "My house in the great metropolis of Oley." Mr Woodward, a history teacher at Hill, wrote, "We attended a family reunion in Attleboro, Mass. In an old barn. There were 88 of us and the noble bird was served – 12 of them."

Question: If you could ask your parents any one question and be guaranteed a truthful response, what would you ask?

Mr. Ralston, another one of our top math teachers, replied, "I actually answered this question for myself about 8 years ago. When my father was sick, it occurred to me that there were some questions that if they remained unasked, I would regret after he died. Mostly they were questions of theology and faith. It is good that I asked them." Nick Kierkegaard replied, "Am I a good kid?"

Question: If you had to name the craziest thing you ever did in your youth, what would it be?

Mr. Ruth, whom I always kidded about being so old, replied, "The craziest thing that I have done in my youth is to continue to answer these questions that appear in my mailbox on Saturday evenings." Peter Kerchner, another one of the DLSF 6, wrote, "Signed up for Algebra I class my second form year with a guy named Pierre." I should add I also had Peter for Algebra II and pre-Calculus. Robert Liou, a DLSF, wrote, "Have Mr. Pierre as a teacher." Again I must include another answer from Kate Carella. She wrote, "When I was about 7 I tried to 'save the ants.' We had a few cement stairs leading up to our door and there were lots of little cracks in them. Hundreds of ants would go in and out and I would watch them and name them (during my tomboy stage). Then one day my mother said that she wanted to get rid of the ants and that she was going to pour boiling water on the steps. So I ran to my room and got a shoebox when my mom wasn't there and I put as many ants as possible in the box…. and then carefully placed it in the back of my bedroom closet. A few days later, however, my mom noticed the steady stream of ants coming out of my doorway! She wasn't too happy about the exterminator's bill." Amish Patel, a DLSF, wrote, "I refuse to answer this question because I know whatever I say may and will be used against me." Conyers Davis wrote the following from St. Andrews, "Join the *SNS*."

Question: If you had to name the one thing that repeatedly makes you angriest, what would it be?

Nick McCartin, another DLSF whom I never had in class, answered, "When my parents say no." I can relate to that answer, which is why it is included in this book. Fred Bobb, who seldom fails to answer an *SNS* question, wrote, "Every six months… spring forward – fall back. Congress needs to repeal this outdated waste of a ritual." Seth Sharp, a prefect in a second form dormitory, answered, "All of the second formers in my dorm who do not know what study hall means even though it is now April." Chris Clarke, a DLSF in my algebra class at the time wrote, "Tests (Hint-Hint)"

Question: If you could proclaim a new national holiday, what would it be and how would we celebrate it?

DLSF Peter Kerchner, a lefty, wrote, "There already is one but nobody knows about it: Left-handed Peoples' Day." Kate Carella (again) wrote, "The new national holiday I propose is National Physics Day in honor of Sir Isaac Newton. It would be on April 12 because that is about the time when apples grow on trees. It was an apple that dropped on his head during the 1600s and prompted him to understand gravity." I believe she had just written a school report on this great man. David Zuckerman wrote, "DLSF Day." Charlie Evans replied, "Oct. 21, of course, my birthday. People can celebrate it any way they wish." Amish

Patel, again making fun of his favorite teacher wrote, "The National Holiday I would declare is Ruth Day. On this day everyone would have to sit in front of their computers and correct term papers, and in their free time they would have to watch CNN." Steve Tsai answered, "National Procrastination Day – we would celebrate it by putting it off until the next day." Damien Newton wrote, "I would have a holiday after America's greatest president, Mr. James K. Polk. Like Mr. Polk, we could use this holiday to get all of our work done on time. I could think of no better way to celebrate the holiday of America's only successful one term president." It should be noted that Damien honestly believed what he wrote. Seth Sharp, a baseball player as well as a prefect, wrote what to me was a great answer, "National Baseball Day--no work, all Americans should attend some sort of baseball game whether it is professional or little league. The typical food served that day would be ballpark cuisine. I would make it opening day of each season. All students would have off from school if they wanted to go to the ballpark."

Question: What is your favorite clean joke?

Rich Voliva, a DLSF, answered, "I can't think of one." William Bouvel wrote, "That you are only 39." I always insisted that I was 39. (Of course few current students have ever heard of Jack Benny.) Mr. Pentz, another one of our illustrious math teachers whom I have used as a crutch on many occasions, answered, "Ford motor company has a vast recall of all the Mercury cars manufactured in the last nine years. No kidding! Why? Traces of tuna have been found in all of them." Seth Sharp, who was not only a prefect and a baseball player but also a DLSF in my Algebra I class, wrote, "Three guys walk into a bar, the fourth one ducks." Tim Kelly responded with "I told the ticket lady, send one of my bags to New York, send one to Los Angeles, and send one to Miami. She said we can't do that. I said, you did it last week." I suggest, how many DLSFs does it take to change a light bulb? Ten, one to hold the bulb and nine to turn the ladder.

Question: If you could sit and have a beer with three sports figures from any time, whom would you pick?

Mr. Kelly again, "Three of the great drinkers according to lots of rumors and some fact were Babe Ruth, Mickey Mantle, and Mickey Lolich. Of course, I wouldn't want to pay the bar bill-- one of the four of us may drink a lot of beer." Matt Holt answered, "Don Mattingly, Mantle, and DiMaggio (the man was married to Marilyn Monroe)." Fred Bobb replied, "Been there! Done that! Bobby Bonilla--NY Mets at the time--in Ruby Tuesdays, Port St. Lucie, FL, ... Pete Sampras in Tampa, FL and at the Hershey Hotel in Philadelphia, PA, ... Pete Rose then manager of the Cincinnati Reds."

Question: Last weekend we moved our clocks back one hour. How did you use your extra hour?

Damien Newton wrote, "I remembered that we had the extra hour but knew the others in our play wouldn't (and knew they would be sleeping in despite a 10:00 rehearsal the following morning) so I called them and pretended that rehearsal had just started and went back to bed. By the time I and the rest of the cast showed up, they had been there for half an hour." DLSF Rich Voliva answered, "I spent the extra hour playing Risk." X Prines wrote, "Studying for an exam while on a band trip." Janak Vidyarthi answered, "We had an extra hour? I must have slept through that." Peter Kerchner, a DLSF going to college in Boston, answered, "I spent the extra hour trying to figure out why The Hill never sends me invitations to events hosted here in Boston, but they sure as hell know exactly where to find me when they want money." Jim Hollister, a DLSF, wrote, "I got to work an extra hour at work and didn't get paid for it. It was just DUCKY!" Lucian Deaton wrote, "Spent an extra hour with friends on the hall. Then I went to breakfast early because I forgot to set my alarm back." Lucian, believe or not, is not a DLSF. Philip Gallagher wrote, "Do you realize that Indiana does not change its time like most of the rest of the country? I was in Indianapolis so I did not get an extra hour." Pat Gavin answered, "Some things just shouldn't be shared." Seth Sharp wrote, "I studied for my math test, but even that was not enough to keep me from failing."

Question: What makes a good roommate?

Joe Fitzgerald, DLSF extraordinaire, wrote, "Not having one. I live in a single." Brian Craighill, a DLSF 6 who always gives thought-provoking answers, replied, "Of course all the simplistic, trivial and stupid arguments will happen. So if you put those aside, as well as some privacy, and can still enjoy the company of that person, you have yourself a good roommate. This person may also take you home for a few weekends out of the year, if you are nice. (Thanks Casey.)" Scott Detar, a DLSF in his first year of college, answered, "Let me first answer by saying that my roommate is different. He is gay. Now I have no problem with it except when he brings guys back to the room. I feel that I should not bring girls back as a matter of respect. Why can't he feel the same way? All of you who are still at Hill, learn that you will need to respect your roommate." Janak Vidyarthi, a true baseball fan, answered, "A good roommate will not be a Yankee fan. A great roommate will be a Baltimore Orioles fan." Seth Sharp knew how a roommate can be useful. He wrote, "A person who is quiet, neat and is in a class higher than me so I can get help."

Question: Who is your hero?

JC Groon, a DLSF, wrote, "I have no heroes, just men and women who provide me with superior role models." Nick Kierkegaard wrote, "My heroes are my parents. They are there through thick and thin. They are always there to look after me and my interests." Kevin Hyde wrote, "My hero is my grandfather: throughout his battle with cancer, he was always optimistic about living to see the next day and the next year. I only hope that I can be as brave when facing death." Matt Holt, one of our more thought-provoking *SNS*ers, responded, "All the Presidents and all the Presidents' men, crooked or otherwise."

Question: If you could have any one super power, what would it be?

Joe Fitzgerald answered, "The ability to fly with x-ray vision being a close second." Peter Boxx wrote, "I think I would want my super power to be able to be perfect at anything I did." Ryan Yerger answered, "Read people's minds." Only Sam Carella could have written, "I wouldn't want any. I think having a super power is like cheating." X Prines wrote, "Immortality." Mr. Weaver, former Hill School math teacher, wrote, "The ability to stop fights and feuds." Fred Bobb answered, "I already have a super power and it's the only one you will ever need--The Lord Jesus Christ." Ed Strapp, former Hill student and then athletic trainer at the School, wrote, "The ability to understand women."

Question: If you were to describe an act of true loyalty that you have witnessed, what would it be, who performed it?

Caren Beck answered, "That would be my dog Jasper. Every time you tell him to sit, he sits. Who else does what you say? That is loyalty!" Courtney Steltz wrote, "My friend sticking by the Philadelphia teams, bad season after bad season." Rob Parker, a DLSF and faculty brat, wrote, "My dog Tib used to get excited every time I came home." Erin Romig played on an intramural softball team of which I was the supervisor. She wrote, "It has to be the varsity coed softball team's ability to show up every day on the field ready to perfect their skills. Every one of those players out there was a dedicated athlete." I doubt very much that

Fred Bobb knew in what high esteem I hold Colonel Oliver L. North when he wrote, "Our whole country witnessed Oliver North and his loyalty to the President and our amazing U.S. military. I was much moved." Cam Greenlee, a DLSF, wrote, "My ex-girlfriend going out with one of my best friends." Leighton Brown answered, "No doubt, my mother when my father was deteriorating due to ALS." Beckie Knight, a student in my Algebra I class, wrote, "When my brother and I were little, he and his friends always created these clubs, and they would always be for boys only. But he would convince them that I was the best tree climber and could fit best under cars to look for balls. So he convinced them to let me, a girl, join." Tommy Manzo, math student extraordinaire, wrote, "My dog Snickers certainly is loyal. This one time I was shooting on my lacrosse net, and she tried to catch my shot in her mouth and got cracked full-speed in the side of the head. It didn't even faze her. She gave out a quick yelp and then immediately retrieved the lacrosse ball and brought it back to me. Now that's loyal." Damien Newton wrote about two of his classmates. J-ball is javelin ball, a game something like baseball, but played with a tennis racket and tennis ball. During the Spring Term Hill School students run a J-ball league in the evenings after dinner. Damien wrote, "Former athletes William Yinger and Rich Voliva chose to remain on my 'Fighting Banana Slugs' J-ball team instead of taking offers to be on teams that didn't lose all their games."

Question: If you had to name the one thing that most frightens you about growing old, what would it be?

Luke LaBranche wrote, "Having someone else helping me go potty." DLSF Mike Weigley wrote, "No more free money from Mom and Dad." I can't wait until Tommy Manzo's father reads his son's answer to this question. Tommy wrote, "Acting like my Dad." Jon Burklund wrote, "Spending my Saturday nights answering surveys." Art Stetson, a Hill graduate who returned to be on the faculty for a few years, wrote, "Losing my friends and family one by one." Shawn Fernandes wrote, "Wrinkles!" (I guess he did not mind losing his sprinter's speed.) Bill Barnshaw replied, "Having to grow up." DLSF Paul Weaver wrote, "Having to act like an adult." Brian Jackson wrote, "I may be corrupted into thinking that 0.99999 … is equal to 1." I proved this fact for his class at least five different ways. I do not understand why Brian cannot see equality as absolute fact. He probably has never heard of Virginia O'Hanlon and so he does not even believe in Santa Claus either. Nick Gerasimowicz simply wrote, "Impotence" Conyers Davis, answering from St. Andrews, wrote, "Not fulfilling my aspirations." Adam Tagert wrote, "Not checking my e-mail enough."

Question: If you could have had any number of siblings, how many would you have had and of what gender?

Mark Piccirillo answered, "One little brother so I could beat him up." Mr. Weaver wrote, "12, gender not an issue. That way I could spend a month visiting each one and not need a house. I would love to have a lot of sisters around my age. Growing up, there is nothing better for a guy, than the friends of his sisters." Mrs. Bedell, mother of a Hill graduate and herself a librarian at The Hill, wrote, "Would have been nice to have one brother to prepare me for living with three sons." Taylor Handwerk answered, "Zero, because I like being spoiled and I hate sharing." This answer shocked me because Taylor is anything but a spoiled child. He is an impressive young man. Julie Horvath wrote, (again knowing her mother would read the answer) "Well, I already have one brother; can I just give him back? I think two brothers and one sister might have made for an interesting time. Please add the following disclaimer: Mom--this does not mean you are getting four grand kids out of me." AJ Farman answered, "3 brothers so that football teams in the backyard would be even." Ben Holskin answered, "I would want one brother and one sister, but I'd still want to be the oldest; not getting beat up by an older sibling is definitely a plus." Sam Carella wrote, "I have the coolest brother and sister in the world, don't need any more. I would have one brother ... one sister... two step sisters... and two half sisters... and if possible their names would be John, Kate, Sue, Karen, Emily, and Victoria respectively." Philip Gallagher again tried to get a stir from his mother, writing, "I'd want four brothers, so that we could have a basketball team (and perhaps a sister to bring us water)." Mrs. Gallagher, Philip's mother, wrote, "I like family and wish that mine were larger ... I think I would have been happy with one sister and two brothers and my cousin's family could have had the same amount and then when holidays came around there would be more than 5 people at the table." I should add that I am the cousin of Mrs. Gallagher. Mr Gizzi, the head football coach and girls' basketball coach, wrote, "Girls starting five and an offensive line..."

Question: If you had to pick the member of your family who is least like the others, who would it be?

Nils Groten, a DLSF, wrote, "Hector, because he is a cat." John Carella answered "Definitely my brother Sam. He's got a way of his own." Brian Jackson wrote, "I would have to say my sister. She does not have as much common sense as the rest of the family." Brian Craighill answered, "My brother..... hands down." Philip Gallagher again tried to get a reaction from his mother. He wrote, "My mom--she's crazy--The rest of us are pretty sane." Cam Greenlee replied, "Me... the black sheep, or so my mom says." Alex Blood sent in, "My little brother. He is such a trouble maker." Again, Amish Patel wrote about his favorite teacher: "Since I consider all of The Hill School to be my family, I'd say that Mr. Ruth was least like the rest." Charlie Evans wrote about his much younger brother, stating, "I think my little brother is least like any of us. He is funnier than any of us and gets more girls than any of us could hope to." Mr. Kelly used some illogical thinking and rationalization to write the following, "Since most members of my family are different, I think that I will choose myself, because I am the only normal one." You would think that as a math teacher, he would be more logical. Jamie Knise, a DLSF who has a very high opinion of his mom, as he should, wrote, "My Mom because apparently she was a good kid growing up and no one else in my family was or is."

Question: If you had to name the nicest thing that has ever happened to you completely by chance, what would it be?

Patrick Lundquist, a student in my Algebra I class, wrote, "I don't know... I'm a pretty lucky guy... of course... I got put in Mr. Pierre's math class... sometimes my good luck just runs out." I could fill an entire book with Patrick anecdotes. Maybe I should threaten him with that. Rob Parker has the ability to make very short sentences say so much. He simply penned, "Once, a girl looked at me and smiled." Poor Robert Liou tried to make light out of the sentence he served while a student at The Hill. He explained, "The opportunity to live on Mr. Ruth's hall three straight years." Tom Power wrote, "This lady came into my law office and said she wanted me to sue McDonald's on her behalf because she had spilled coffee on her leg in the drive through." Again Courtney Steltz wrote a poignant reply, "Ending up with Mr. Drowne as my adviser at The Hill. It would have been nearly impossible without him." Clay Shaner wrote, "Several years ago I was sitting with friends at a restaurant. We were trying to figure out what we would do for summer jobs in order to make enough money to do more fun things. When we asked for our bill, the waitress said it had already been paid by the gentleman who earlier had been sitting by himself next to our table. Unfortunately he had left before any of us really got a look at him or had a chance to thank him." This brought back some vivid memories to me as I was on the other end of a very similar situation. Sitting in a diner I overheard the two young kids in the next booth talking. They realized they did not have enough money to pay their bill because they had forgotten to figure in tax and tip. As I was leaving I asked the waitress if I could just pay their bill for them. I did so. As I drove out of the parking lot the two kids came outside, running and waving and all smiles. Seeing their reaction made my day.

The weekend before the elections in 2000 *SNS*ers were asked to vote for president. *Survey Says* did not list each response but gave the following summary.

"*SNS* declares George W. Bush the winner. In a very close race members of *SNS* voted in favor of Bush. Here are the breakdowns of the final tally.

Popular vote: Bush 53   Gore 52
                         Nader 11   TRuth 4

Other responses:
   Hate both of them
   Brendan McGowan
   Undecided
   None of the above

Current Hill students – Bush 4   Gore 4
Current College students   Bush 14   Gore 20
Post college (age >22)   Bush 35   Gore 28
DLSFs   Bush 9   Gore 12"

One day a colleague asked me an interesting question. I asked if I could use it for *SNS*. He readily agreed. You are a member of a very small church group with a congregation of approximately 100 parishioners. Your minister has taken out of her own pocket approximately $100 and gives a one-dollar bill to each member. You can do with it as you wish. Hopefully it will bear fruit or you can get it to increase three-fold or more. What will you decide to do with your dollar?

Brian Craighill wrote, "I would give the dollar back to the church." The ever thoughtful Mrs. Dollhopf wrote, "I would buy a flower and give it to someone who was going through a rough time. Enclosed with the flower would be a note. I would explain about the dollar, how I wished to use it to spread happiness, and how I hoped this person would do the same. In this way a chain of happiness might be formed." Krista Andersen replied, "Buy food for a hungry person on the street." Erin Romig answered, "I would probably buy some bubble gum or a lottery ticket." Ryan Yerger answered, "I'd buy a big candy bar, maybe a fund raiser candy bar. Then the money would do some good." The creative Kate Carella explained, "I'd buy a package of colored construction paper, use my crayons and create greeting cards to sell to my friends and acquaintances. Then I'd return the dollar along with some extra funds to the church and brighten someone's day with a cute card in the process." Bill Barnshaw responded, "Put it in the offering plate." Jim Hollister wrote, "Place it in an interest bearing account and weekly add another dollar. After fifty years I would close the account and give it back to the church." I can see some bank now receiving all these one dollar deposits to be left for fifty years. Mrs. O'Shaughnessy, a long time faculty member at The Hill, responded, "Well, since a minister gave it to me, I don't think I can justify buying something for myself, and what can you buy for a dollar anyway. Hmmm. I could give it to a grandchild and watch the delight on his/her face. Or, I could buy four mint state quarters and hope they increase in value some day. Where do you get these questions?" Brendan McGowan missed the point of this question, I believe. He wrote, "I would buy cheap shoe laces at Target, I need those." Melissa Reichenbach, a Hill graduate and currently a cadet at the Naval Academy, certainly wanted to help many others as she wrote, "I would buy a flower for the altar the next Sunday to make people smile. Everyone smiles when they see flowers." Paul Fulmer had a unique suggestion, "I would ask Hillary Clinton to show me how to invest it in cattle futures." I have no idea what Mike Weigley, a DLSF, was thinking when he wrote, "I would buy a ball of string." Sam Carella obviously knew what three-fold meant. He answered, "Fold it over three times and give it back." Mrs. Kelly, in one of her shorter responses, stated, "Sometimes stores sell packets of seed at the end of the season for only ten cents a packet. I would buy ten packets of assorted vegetable seeds, plant them the following spring, and donate the resulting produce to the church or a local food bank for distribution to those who need it (My dollar would rather literally "bear fruit"!) The value of the produce would exceed by far the small financial investment, and my labor in the garden is always time well spent." Julie Horvath obviously was thinking along the same lines as her mother, but was able to say it in many fewer words. I am sure that she did, once again, make her mother proud. She wrote, "Here's an answer to make my Mom proud... I would buy some seeds and plant a garden to help others who are less fortunate." Mrs. McCallum does receive *Survey Says* so she eventually got all the answers she was looking for. Her reply was, "If I could figure that out I would be a multi-millionaire--Need answer immediately."

The Hill School's decision to become a coeducational school did not sit well with many students and graduates. When the decision was first announced there was quite an uproar as it angered many constituents. I could not help taking advantage of the situation by playing mind games with SNSers by asking the following question.

Question: Since The Hill School is going coed, what do you think the new School colors should be?

Ben Walborn, a DLSF, replied, "I think that this idea is a bunch of crap. Keep the colors." Daryl Groon wrote, "Pink like the liberals who decided to let the school go coed." Mr. Drowne, the real one, wrote, "The colors should stay as they are.... Period." Kevin Miller wrote, "Do not change the colors. What's next, no sit down dinners, no dress code, dropping the athletic program, moving the school to New Jersey? Come on people, maybe there is a need for change, but let's not go too far." Henry Clifford vehemently suggested, "Never, ever, ever, ever should the school colors be altered from blue and gray to anything else!

There is so much tradition in a place like Hill. Need we trash it because the School is admitting women?" Peter Kerchner showed he was upset by writing, "Changing the school colors goes a little too far, not that they haven't gone too far already. This is getting more and more offensive every day." Shawn Fernandes sent in his plea, "Please do not try to destroy The Hill any more than has already been done. Isn't anything sacred?" Mrs. O'Shaughnessy showed her concern for the football team: "Are we going to have a football team dressed pink and purple? And knowing some women, myself included, they would be distressed to think that their gender had a color attached to it." For once Amish Patel did not include Mr. Ruth in his answer. He wrote, "I think the colors should be puke green with a pinch of purple mixed in with gray polka dots. And everyone must wear fluorescent pink and yellow Irish kilts." I found it interesting that so many were concerned about the athletic uniforms. Mr. Kelly, a football coach, considered a strategic advantage. He wrote, "I vote for maroon and brown because they hide the ball better on options." Scott McCallum, son of Mrs. McCallum and a DLSF himself, had a great reply: "Pink and yellow and how about a handful of daisies for the new crest."

Question: What would you place in the time capsule to be behind the cornerstone in our new Academic Center?

Zach Brusko replied, "A picture of this year's senior class, the last all-male class." Lucian Deaton suggested, "This year's yearbook. It'll be the last not to have girls in it." Peter Abrams, one of the graduates most upset by the coeducation decision, wrote, "I would place a little miniature of The Hill School campus--a symbol of a school that was once distinctive and in its own way extraordinary, but sadly no longer exists." Rich Voliva wrote, "I would place stuff that had to do with the all-male identity of the School. After all, that will probably be forgotten in ten years." Henry Clifford obviously had tongue in cheek as he replied, "I would place a bowl of The Hill's yummy Seafood

Newberg to see what effects time has on the yummy stuff." X Prines answered, "A tie, a man's dress shirt, a Hill blazer, and pants--just in case the School finds the need to do away with all male things."

I have not included the editorial comments I often made to answers because most of them were inside jokes. *Saturday Night Survey* is going stronger than ever today. I just finished typing this week's answers to what is your favorite poem? I was gratified to see many members submitted their own poetry. Any reader of this book is welcome to join *SNS*. To become a member of *SNS* all you need to do is send me your e-mail address.

Chapter VIII

Fibonacci

Sometimes a teacher finishes a lesson a few minutes early and needs something worthwhile to conclude the class, preferably something that provides an exciting learning experience and at the same time arouses curiosity in the students' mind. One of the topics I found interesting and useful is the Fibonacci sequence. I broke the Fibonacci sequence down into small pieces which I used for days and days. It never ceased to surprise students to find these numbers showing up in so many places.

Leonardo Fibonacci wrote *Liber Abacci,* one of the earliest books to define the symbols for arithmetic. It was also loaded with exciting mathematical problems. One of the problems has become known as the rabbit problem.

A pair of rabbits produces a pair of rabbits every month. However, a pair of rabbits must live for two full months before they can start producing. If a new born pair of rabbits is put in a pen on day-one of month-one, how many pairs of rabbits are in the pen after one month, two months, etc?

This problem is the perfect introduction to the Fibonacci numbers. At first, I never mentioned Fibonacci in class. I just gave the students this problem to solve together. They were required to guess the number of pairs of rabbits in the pen at the end of each month. I usually had to label the pairs (A, B, C, etc.) to help them discover the sequence of answers. I drew and then filled in the chart as the students came up with the correct answer for each month.

Month	Pairs	Names
1	1	A
2	1	A
3	2	AB
4	3	ABC
5	5	ABCDE
6	8	ABCDEFGH
7	13	ABCDEFGHIJKLM
8	21	ABCDEFGHIJKLMNOPQRSTU
9	34	
10	55	
11	89	
12	144	

Usually, after a very short time, at least one student saw the pattern and was able to finish the chart. Then that student explained the pattern to the rest of the class. After the table was complete through twelve months, I stopped and gave some history about where this problem first appeared. Fibonacci knew this sequence only as the solution to his rabbit problem. Everything we explored about these numbers had been discovered by Fibonacci's successors. Most years I would have time on that first day only to introduce the rabbit problem and give a brief history of its origin.

The next time I worked with the Fibonacci numbers, I wrote them across the chalkboard and numbered each term.

Fibonacci number	1	1	2	3	5	8	13	21	34	55	89
term	1	2	3	4	5	6	7	8	9	10	11

Fibonacci number	144	233	377	610	987	1597	2584
term	12	13	14	15	16	17	18

I asked the class which of these numbers are divisible by 2. Yes, every third Fibonacci number is divisible by two and two happens to be the third Fibonacci number. Which Fibonacci numbers are divisible by three, by five, by eight etc? Students were always fascinated that such an interesting phenomenon occurred.

Usually on another day, I explained that the Fibonacci numbers are my favorite numbers. However, one number in particular stands out as the most significant. That is the number 89. I then explained that my favorite number is not actually 89, but the reciprocal of 89. This observation usually warranted a review of the reciprocals of numbers. I then told the class that I wanted the reciprocal of 89 written as a decimal number. Most students knew to divide 1 by 89. I insisted that the division be done without a calculator.

After the students had seen that the quotient has the digits 011235, I asked them to guess the next digit. Every student at this point always guessed 8. I then showed the next digit was a 9. Then I continued to do the division to get the sequence in the quotient as 011235955.... But I explained that I could live in my own little fantasy world if I wanted and could believe we had an 8, not a 9, as the seventh digit in our quotient.

I wrote on the chalkboard .0112358 and asked what should come next. Every student knew it should be 13. Since 13 cannot be written as a one-digit number, I wrote the sequence up through 8 and put the 13 underneath. .

```
    .0112358
       13
```

... 3 and one to carry (as every grade school student learns when doing long multiplication). Next I continued to write the Fibonacci numbers into the problem on the board.

```
    .0112358

        13

        21

         34

          55

          89
```

```
    .011235955.........
```

After we had added the terms as we would in long multiplication, the students saw the result was the terms of the Fibonacci sequence. The base-ten number system kept us from recognizing the Fibonacci sequence. This reciprocal of 89 problem always led into a discussion of other number systems using bases other than 10. In the discussion the students explored the advantages and disadvantages of our common base-ten number system. Also, I always made sure that we at least touched on non-base number systems.

Probability is not in many high school curricula. I believe students should have a basic understanding of probability because it is a powerful tool. The probability of an event's happening is simply the number of favorable ways it can happen divided by the total number of ways it can happen. So the probability of tossing a fair coin and landing heads is ½. There are only two ways the coin can land and there is only one favorable way.

After the students were comfortable with the definition of probability I posed the following series of questions. What would happen if we tossed that coin twice? What is the probability that we would not get heads both times? What if we tossed that coin three times or four times, what is the probability that we would not get two heads in consecutive tosses? Since most of the students had never worked with probability before, I listed all the outcomes.

Toss Twice	Toss Thrice	Toss Four Times
HH	HHH	HHHH
HT	HHT	HHHT
TH	HTH	HHTH

TT	THH	HTHH
	TTH	THHH
	THT	HHTT
	HTT	HTHT
	TTT	HTTH
		TTHH
		THTH
		THHT
		TTTH
		TTHT
		THTT
		HTTT
		TTTT

The students counted the total number and the favorable number for each situation. The fraction of the favorable ways divided by the total number of ways was the probability.

Twice	Thrice	Four Times
3/4	5/8	8/16

Every student quickly saw the numerator is always a Fibonacci number and the denominator is always a power of 2. If time permitted, we continued to find the probability for a larger number of tosses.

The Divine Ratio or Golden Mean goes back to the days of the early Greeks. They considered the Golden Mean to be the ratio of two quantities that was the most pleasing to the eye. Most Greek buildings were constructed with this ratio in mind. The exact value of this ratio is the quantity 1 plus the square root of 5 all divided by 2. I had the students punch this ratio into a calculator. The result approximated 1.618. The students then had to look at the ratio of two successive Fibonacci numbers.

$$3/2 = 1.5 \quad 5/3 = 1.667 \quad 8/5 = 1.6 \quad 13/8 = 1.625$$

$$21/13 = 1.6154 \quad 34/21 = 1.619 \quad \text{etc.}$$

The ratio of two consecutive Fibonacci numbers gives a good approximation of the Golden Mean, and the larger the two consecutive Fibonacci numbers, the closer the approximation is to the Golden Mean.

Students do all math work in base-ten and therefore find it difficult to visualize and work with systems in other bases. At this point I introduced a pseudo-base system. The base-ten number system has the units' place being the number of tens to the zero power. The tens' place is the number of tens to the first power. The hundreds' place is the number of tens to the second power. So the number 123 means 1 times 10 squared plus 2 times 10 to the first plus 3 times ten to the zero.

I introduced my students to my pseudo-base number system. The reciprocal of 89 problem was not for naught. We used instead the Fibonacci numbers as our base. That is, the first place would be the number of ones, the second place the number of twos, the third place the number of threes, the fourth place the number of fives etc. (For this system we use only one of the ones from the Fibonacci sequence.)

Therefore the number one in base-Fibonacci is still 1. The number two base-ten could be either 2 or 10 in base-Fibonacci since it could be read as two ones or one two and zero ones. I stated that I had made a rule that the system could use only the binary digits zero and one. There cannot be two different ways to write the same number! Therefore, two base-ten would have to be written 10 base-Fibonacci. Three base-ten could be 100 or 11 in base-Fibonacci. Again, there never should be two ways of writing the same number! Just as was done with order of operation, an agreement solved this problem: never have two ones next to each other. The following chart shows the base-ten numbers with the corresponding base-Fibonacci number.

Base Ten	Base Fibonacci
1	1
2	10
3	100
4	101
5	1000
6	1001
7	1010
8	10000

9	10001
10	10010
11	10100
12	10101
13	100000
14	100001
15	100010

This pattern continues and forms a perfectly good new number system. In fact, this number system is better than our base-ten system for solving probability problems similar to the consecutive heads coin toss problem. But what is a book about mathematics that does not have that famous one sentence. "That is beyond the scope of this book." My other favorite expression that seems to appear in most math books is "the proof is rather trivial and is left to the reader." I always have difficulty doing those proofs.

Most of my students had favorite numbers so they went along with mine the reciprocal of 89. However, they found it difficult to understand that I would also have a favorite function. I told my kids my favorite function is

$$f(x) = \frac{1}{\sqrt{5}} \left(\left(\frac{1+\sqrt{5}}{2} \right)^x - \left(\frac{1-\sqrt{5}}{2} \right)^x \right)$$

As part of an assignment I always had the students evaluate f(1), f(2), f(3), f(4), f(5). Likewise, I will leave this problem for the reader to solve.

The Alumni Chapel at The Hill School has a row of stained glass windows, each devoted to an outstanding person in the field depicted on that particular window. I am always pleased to look at the window for Arithmetic as it is devoted to Fibonacci. I must include my two other favorite windows. The chapel has no window for calculus, a subject area discovered by Sir Isaac Newton. Newton belongs among this elite group and is honored in the science window. The window for geometry seems to amaze many people. Eudoxes preceded Euclid. Some people think that a good part of Euclid's work in *The Elements* was taken directly from Eudoxes. Obviously, Euclid's work was better organized and has lived through the centuries. However, the geometry window at The Hill School recognizes Eudoxes. Unfortunately, there is no window to honor the Great Archimedes. His contributions in the field of mathematics must rank him as one of the great mathematicians of all-time. In this writer's mind Archimedes was the greatest mathematician of all-time. One of my students once asked if I thought Archimedes' first name was Great. He stated that I never said Archimedes. I always referred to him as the Great Archimedes.

Chapter VIIII

Reminiscences

I treasure so many little incidences that do not seem to fit elsewhere in the story. In this chapter I want to include them because they were an important part of my teaching career. Here they are listed, in no particular order.

Jonathan Clarke is a remarkable young man. I first met him during summer school before he entered his third form year at The Hill. He told me that he knew he never would have been accepted by The Hill School except that he had an older brother who had been very successful at The Hill and a father who was a Hill School graduate. The next day I went into the office to check his file. I sat down with him that night to explain that he had been accepted on his own merit. There was no question that he came to us with great recommendations and a solid background. Towards the end of his first year he got himself into a little difficulty. I explained to him that I would like to take him out to dinner and have a good chat about his situation. At dinner, I was impressed that he fully understood what he had done wrong and was quite remorseful. I will never forget what he said when I returned him to his dormitory that evening. "Thank you very much for dinner, but more importantly, thank you for taking an interest in me."

I had never known anyone before who never answered a question with just a "yes" or "no." If I were to ask, "Is your name Jonathan Clarke?" His response would be, "It is." Every answer was a complete sentence and a solid response. I enjoyed this trait so much that I often asked him a question that could easily be answered with a yes or no just so I could hear his response. I never mentioned this idiosyncrasy to him because I thought he would be self-conscious about it and try to change. I did not want to see Jonathan try to refrain from these answers as I thought it was an admirable trait. It would be interesting to have Jonathan moderate the old radio show *Twenty Questions*. If Jonathan reads this book, I hope he does not try to change this trait.

One year a DLSF in my Algebra I class and I made a bet on a game of tennis. If he won, I would treat him to a steak dinner. If I won, I would get to choose his workjob for the following year. It so happened that I was in charge of the breakfast workjobs. This young man had to arrive every morning ten minutes before breakfast to put plates and water on the table and remain after breakfast to set the table for lunch. He did not enjoy having to get up that early, especially knowing that many of his classmates were getting an extra fifteen minutes sleep. We played our tennis match, and when I won, I informed him that he would be setting table number four for another year. Brian Craighill came to breakfast early and did his work job every day that second year and never once complained about it. I was even able to convince him to work yet another year in the dining room, but as a captain who had to make sure others did their jobs. I still marvel how he did his workjob every day a second year without a complaint. There are very few young men who could have done that. At such a young age he too was a man of his word.

One day I was informed by the mother of one of our students that her son was in a musical production in Reading. I bugged the student periodically until he finally gave me the information so that I could get tickets to see his performance. I called to order the tickets and was told that I could pick them up at the door the night of the performance. It was a thoroughly enjoyable performance, Patrick Lundquist as Jean Valjean in *Les Miserables*. What was even more amazing was that he obviously had arranged very special seats. We sat front row center, the best seats in the house. He has yet to mention how he managed to get us these special seats, and whenever I asked him, he changed the topic of conversation.

Patrick Lundquist is also one of the young men I asked to help me move in June of my last year at The Hill School. I told each student who worked that I would pay him. I know that Patrick did twice as much work as any other student. He always took the heavy loads and did not stop often for breaks. When all the work was complete and I handed each boy an envelope with his money, Patrick replied that he had not done this chore for money and refused to take the envelope. I insisted that he take it. He did, but later that night as I got into my car, there on the front seat was Patrick's envelope with the money still in it.

I was at a Hill football game one Saturday, taking photos during warm-ups and the opening coin flip. As the game was about to begin, I walked up into the stands to talk with some friends whom I had just seen arrive. As the National Anthem began, I did a double take. Out there on the field was Patrick Lundquist in his football uniform singing the National Anthem. I took a few photos from the stands, but the ones I would like to have taken, if I had known he would be singing, were from field level. Everyone in the stands marveled at the beautiful job he had done. When I saw Patrick the next day, I asked why he had never said anything to me as I would have stayed on the field and gotten some really great shots. He replied, "It just was no big deal." I am sure to him it wasn't, but to everyone else at the game that day it was a big deal.

Often when sending notes to parents I would place them in their child's mailbox. I added notes to the envelope with directions to the student for delivery of the envelope. I wanted to send Patrick Lundquist's parents a message. I addressed the envelope to the parents and added a note, "Not to be opened by Patrick." Another teacher was in the mailroom the day I placed the envelope in Patrick's mailbox. The teacher stated that he thought there might be five students in the entire school who would actually take envelopes home to their parents without opening them. I agreed, but I said, "Patrick is one of those five." The telephone operator spoke up, "Mr. Pierre is right--Patrick is a great kid." How, I can only wonder, does a telephone operator get to know students well enough to make such good judgments about them. For the reader's information, he did not open it.

One day, Taylor Handwerk, a sixth former, showed up about fifteen minutes late to take a test in Calculus class. I handed him his test and watched him begin to work. Perhaps five minutes later I wrote a note on a scrap of paper and handed it to him. "Is something wrong? You look like you are ready to explode!" He nodded and jotted on the bottom of the paper, "Yes, big fight with my girlfriend. It's o.k. I'll be fine in a few hours."

Over the years I have been given many baseball books by students. Probably long after I have forgotten the contents of the books, I will remember the inscriptions written in them. The students thought about the inscriptions, and when they wrote, they gave of themselves. Here are a few of the most memorable.

Ballpark with the inscription "Thanks for being a great teacher," was penned by Derek Giangiulio along with Derek's autograph. He also had the author, Joe Mock, inscribe the book.

"I want to thank you for all the help and guidance that you have given me over the years. You made the whole Hill experience a lot more enjoyable. Your friend, DLSF Brian Craighill" was penned in *Baseball Legends of All Time*.

When Patrick Lundquist arrived on campus, we no longer had a second form. However, we did have upper formers in the School who had started in the second form year. In the book *What Baseball Means To* Me", edited by Curt Smith, Patrick inscribed "Although I never got to be a DLSF, you always treated me as one. Thank you for everything." He then made sure I knew he remembered the quadratic formula as he wrote it out perfectly.

When I was in the Director of Studies Office, I ran a used-book store. At the end of a school year, after all the students had left campus, I collected all the books they left behind. Then, on shelves in my office, I organized the books to be used the following year. Scholarship students could come into the office and pick up any books they needed. Then other students could come in and buy books at twenty-five cents for a paperback and one-dollar for a hard back. Scholarship students used this money when they needed to buy some athletic equipment. Charlie Van Voorhis, a DLSF, was extremely bright and very inquisitive. He would come in to buy paperbacks before vacations so that he would have something to read on his train ride home. Once he came in and found four books he wanted and said that he was broke. I told him that he could have the books for one-half of all the money in his pocket. He fished out three quarters, a dime, and two pennies. I thought to myself, great, an odd number. How will he divide eighty-seven cents in half? Charlie quickly picked up the three quarters saying, "These three for me and those three for you!" as he pushed the dime and two pennies across the table to me. His quick thinking earned him four books for twelve cents.

Josh Poley had very poor handwriting. When he wrote his last name, to me it looked like Pokey. After I had mentioned in class his name looked like Pokey, all his friends began to call him Pokey. I had him in class again. On opening day, he answered roll call "Pokey." He was a day student. Often day students take what some faculty call a "dayboy holiday," a vacation day using a phony excuse. Whenever there was snow, no matter how little snow, Pokey took a "day boy holiday" claiming the snow was much deeper by his house. One day, when we had some snow flurries and he did not come to school, I told the class that we had to do something about his repeated absences. I told the kids that the next day he showed up, I would make something up and when I called on them, they were to give me any answer. Their answers would always be correct, and if Pokey volunteered an answer, it would be wrong. I had just finished reading about Isaac Newton's work with fluxions and the term stuck in my head. So the following day, when Pokey arrived in class, I announced we would review the work we had done with fluxions. I gave 7 prime cross 8 prime and called on someone to give me an answer, "21." I called on someone else to find out if that was the correct answer. Of course, "yes." After about five of these questions I gave 19 prime cross 3 prime and called on Pokey. He evidently figured out we were pulling a fast one and made up an answer. When I called on another boy to see if Pokey was correct, he said "no" of course and corrected with a different number. Finally, at one point, I called on Pokey and after he had given his answer, I called on John Bolger, who was sitting next to him, to see whether Pokey was correct. John said, "Yes." For some reason I went back to Pokey to find out how he had solved the problem. He said "John Bolger whispered the answer to me." After a short time and a few more of these problems, Matt Brewster, who was sitting in the back of the room, said "This stuff is so easy, let's have a quiz on it." I agreed that was a good idea. "Let me explain the new material and then we will have a quiz on fluxions." As I began the explanation, I noticed Pokey turning to the index and trying to find fluxions. When he could not find them, he began to leaf through the book page by page. I guess he wanted to learn how to do those problems before the quiz.

Mike Weigley was one of the fun students I had as an advisee. He was always happy and enjoyed having a good time. He also possessed a quick wit. He had a particularly great way of interacting with adults. Whenever I had guests or visitors on campus and wanted a student to give them a tour or take them to the school dining room for a meal, I always asked Mike. It never failed that I would get great comments afterwards from the people he had escorted. Mike had a great love for Boston Cream doughnuts from Dunkin Doughnuts. He also was usually one of the very first students to walk into the dining room for breakfast each morning. Breakfast was buffet style so students could come and go as they pleased. On his birthday, I set the table where Mike usually sat. At his place, he found a giant sized Boston Cream doughnut, a birthday party horn, and a couple presents. One present was absolutely the ugliest necktie I could find. Mike seemed to enjoy the whole thing and commented to everyone as they came into breakfast about his birthday gifts.

Peter Kerchner is a DLSF I had in Algebra I Honors his first year. Whenever I smiled in class, Peter always remarked, "Look, he is laughing at us." Because this was a particularly good class, we did have great fun. Several years after Peter had graduated, I guess he finally realized I had not been laughing at him all the time. He called because he and Rich Voliva, who I also had had in class, were going to a New York Giants vs. Philadelphia Eagles football game. They had an extra ticket and wanted me to go with them. It was a great day for me. The Giants won so I really let poor Peter have it all day long. He probably was wishing he had never come up with this idea.

The day before a long weekend, we had a very bad ice storm. A note was sent around to every room for teachers to read to their classes. It stated that classes would end at noon and that everyone could leave campus then because the weather was to get much worse and by the next morning students might be stuck on campus. I read the note in a laughing fashion so that it sounded as though I were making the whole thing up. Jan Ebersole, sitting in the front row, finally asked to see the note because he did not believe what I was reading was legitimate. Of course, I let him read it and he told the class it was real. That was fine, but the day before the next vacation the weather again was very bad. So I wrote out a note of my own and gave it to Matt Ralston, who was to deliver the note to my classroom in the middle of the period. He did and I read the note. "The weather is so bad that it would be dangerous for students to leave campus. Therefore the vacation will be delayed for two days. By then it should be safe for students to leave campus." Again Jan insisted on seeing the note--it couldn't say what I had just read! So I gladly showed him the note. To my shock at lunch that day the headmaster again announced that students could leave early because of the weather.

Joe Fitzgerald is one of the best math students I have ever had. He also happened to like the game of tennis. We played a set of tennis. I proposed the wager: I would treat him to a steak dinner if he won; if I won, he would write a 2000-word essay about how I was a better tennis player than he was. Needless to say, he had to write the essay. He actually did write it, and a very humorous essay at that. That essay, of course, was included in his scrapbook when he graduated.

Ben Walborn was a very precocious second former in my Algebra I Honors class. I do not remember what got me started, but I began to dwell on all the things wrong with teenagers. After class, Ben waited around until everyone had left and came up to my desk to say that I could not have been talking about him. I must have been talking about all the other students. Then he informed me that he was only twelve years old.

Brian Craighill was one of the most popular students on The Hill School campus. I did not realize just how popular he was. In his third form year, his second year on campus and my twenty-second year at The Hill, I had an occasion to walk across campus with him to my house. Every person we saw said something to him and called him by name. It was embarrassing as more students knew him than knew me! Finally, after each person had passed by, I would ask Brian how he knew him. It was unbelievable. There was no common denominator. One was in the band with him; several were in classes with him, and some he had met in the student grill; a few he had met when he visited the campus two years earlier as a prospective student. I doubt very much that we just happened to see only people he knew.

109

One year, before The Hill instituted an honors program in mathematics I had an Algebra I class in which most of the very brightest students were assigned. Most of them were DLSFs; a few were third formers. One of the third formers was a young man named Scott Schluter. Although very bright, he always seemed to miss factoring problems that involved the common factor. So whenever we came to such a problem, I called on him. One day he was absent and I did not realize who was missing. I asked the class, and one student replied that it was "Common Factor." He did not even know Scott's name, but he realized that Scott was absent. Thereafter we always called Scott by that nickname. I placed a letter in Scott's mailbox after he had been elected a class officer his sixth form year. He told me it was the shortest letter he ever received. I had written his name and "Congratulations" and had signed it. After Scott graduated, I sent him a letter and addressed it to SC(ott hluter) as his name. It arrived at its destination with no problems.

The Hill School has had a string of great wrestlers and many National Prep School Champions. Arguably the best of all was David Hoffman. I had Dan Hoffman, David's younger brother, in class. Dan is no slouch at wrestling. As I write this memoir he still has one more year of high school and very well may become a National Prep Champ himself. Although he has won many accolades as an outstanding wrestler, everyone compares him with his older brother. Whenever a newspaper has an article about him, it always mentions that he is the brother of Dave Hoffman. So I always had to call him "Dave's brother." At a match last year, as I was sitting in the stands taking photos, a woman from the visiting team walked up and asked me which kid was Dave Hoffman's brother. It struck me funny and I started to laugh. Fortunately, I was able to explain to her why I was laughing. I hope that if Dan reads this book, he will take a close look at the index. On that page he has been placed ahead of his brother Dave.

110

Many sports moments stood out during my time at The Hill School. I could mention many baseball games as we had many championship teams. At one point The Hill won the state title five out of seven years and was runner-up one of the other years. In my opinion, the team that was runner-up was probably the strongest team of the group. However, for me, the most memorable athletic contest was a football game against rival Lawrenceville. In 1976 The Hill was celebrating its 125th year. There was to be a big celebration during Lawrenceville weekend. The two football teams seemed to be quite evenly matched and everyone expected a very close game. Never have I seen a Hill team dominate another team so completely for an entire game. It seemed that everything fell into place that day, and in the end the score was 41-8. It may not have been one of the strongest football teams in School history, but on that Saturday afternoon, it played like the best.

The most memorable individual moment in sports came in a basketball game. Arguably the best player during my tenure at The Hill was Barry Pierce, who later went on to captain the University of Pennsylvania to an Ivy League championship and then into the NCAA basketball tournament. No player could possibly go out and finish his high school career on a higher note than Barry. In his final game with seconds left on the clock, The Hill was down by four points, a certain loss. However, a player took a three point shot, and an opponent mistakenly fouled him in the act of shooting. The ball went swish, all net. So The Hill was one point down with one foul shot. The foul shot was off line, but Barry soared into the air above everyone else and tapped it in for two points and the win as the buzzer sounded. Barry's uniform, number thirty-four, was the only uniform The Hill School ever retired.

I also had the good fortune to see Sam Carella run track for five years for The Hill. As a second former he always finished far behind. Once, when he was running the mile, I took a great photo of him coming down the stretch with a slight lead over the star of the opposing team. The one thing the photo did not show was Sam still had a lap to go. I believe in the last race I saw Sam run, he broke the School record in the half-mile. As mentioned earlier, he won many awards and is now running at Villanova.

When I was in the Director of Studies Office, we had a great group of secretaries who were not only hard working and devoted to The Hill School but also hard players when it came to fun times. The leader was Kathryn Miller, registrar. My secretary was Jeanne Krause, and we also had Pat Fort and Dolores Koren. Our office took care of all academics, including mailing transcripts to colleges. On one occasion I was quite distraught because the School had changed the phone system. There had been a buzzer system such that I would ring the buzzer and Jeanne Krause would come in to see what I wanted. The only time I used it was when I had a parent on the phone. I could buzz Jeanne, write a name on a piece of paper, and she would get me the folder. On many occasions I had parents awed by how much I remembered about their sons without having to go look up information. I knew their grades, the names of all their teachers, and even what their teachers had to say about them. Of course, I never mentioned I had their son's folder on my desk as we spoke. After the new phone system had been installed, since it had no buzzer, I had to dream up a new system to contact Jeanne. I brought in a little battery operated car and, saying nothing to Jeanne, I waited for that first call when I would need a folder. I left the door to my office half-way open. At the proper time, I attached the note to the car and ran it out to Jeanne. With the door ajar the car was able to hit the door to make the turn and arrive at her desk. She was laughing so hard I think it took five minutes to get the folder. Our group became known as "Ye Ole Gang." Today we are all retired, and it has been 18 years since I left that office, but we still get together on a regular basis to laugh about old times. Jeanne passed away a few years ago, but we have added Rose

112

Schoenly, who worked in another office at that time, but spent many hours with us in our office.

Ye Ole Gang added a fictitious student to The Hill School roster. No one ever knew where and when this student would show up. His name was Hugh Bluitt. We kept a file on him and gave him grades. There were always letters showing up from people writing on Hugh's behalf. Many people took part in this prank. I must now admit I wrote a few of those letters. Many faculty members would even write comments to send home to his parents. One time I was caught in the Hugh Bluitt trap. The numbers did not come out right when I was doing the grade statistics, and I could not find out what was wrong. After about an hour of scratching my head I realized that Doc Finn, who taught Hugh Latin, had turned in a grade of "F" for Hugh. With that grade out of the mix, the numbers balanced correctly.

One year on April 1, I dictated a garbled message for Jeanne. Evidently she had tried to listen to it several times and could not decipher it, so she had the others in the office listen to the tape. Then they realized that it was April Fools Day. So Jeanne sat down and wrote me a note and left it on my desk. Since it was written in shorthand, I had no idea what it said.

The second formers were taking a trip to Washington. I knew they were to get to meet James Baker when he was Secretary of State. I asked a DLSF if he would take a baseball and get it autographed by Mr. Baker. Unfortunately, Mr. Baker had to be out of town that day and the second formers were not able to meet him. The young man, though, went to James Baker's office and explained to the secretary that he wanted to get a baseball signed by Mr. Baker for one of his teachers. He asked if she could have Mr. Baker sign the ball and then send it to his teacher. She agreed. A couple weeks later I received the ball in the mail with a very nice note from the secretary telling the whole story. She ended the letter with an apology because she had forgotten to get the boy's name and was unable to tell me who had done this favor for me. I often wonder whether the boy would have given the secretary his correct name, if she had asked. I know that one day in this boy's first year at our School, he went downtown with some of his friends, and a couple of them came back to tell me what had happened. As he was paying for his items, the sales clerk asked if he was from The Hill School. After an affirmative reply, the clerk noted the Trump boy went there and asked, "Do you know him?" The young man replied "sort of." His answer to "What is he like?" was simply "O.K." He paid for his items and then Donnie Trump left the store.

After the graduation ceremonies, The Hill School provides a very nice luncheon on the Quadrangle. The entire faculty and all the graduates and their families can gather there to say final goodbyes. One year, Bill Boyd, who had been a five-year boy, came up to me on the Quad. He said, "I want to thank you for everything." He handed me an envelope and ran off to go home. In that envelope was a card on which he had simply written "Thank you for being my friend." Included with the note were four box seat tickets to the next San Francisco Giants vs. Philadelphia Phillies game at Veterans Stadium. I had a very similar experience with a young man named Matt Cecil, who gave me tickets to a San Francisco Giants vs. New York Mets game at Shea Stadium. At that game we sat directly behind Duke Snider.

When The Hill began construction of the new Academic Center, I thought we should have a time capsule in that building, and on several occasions, I pestered the headmaster about it. He finally said we will have a committee--you are in charge and you select your committee. I thought this committee should include a few students. One student I chose was William Bouvel, a DLSF. What a stroke of luck that was! I think he did more work for that time capsule than all the rest of us put together. He came up with

114

the wording on the plaque where the capsule is imbedded in the wall. It reads

FORSAN ET HAEC OLIM
MEMINISSE IUVABIT

THAT MEN'S ACTIONS MAY NOT IN TIME BE
FORGOTTEN, NOR THAT THINGS GREAT AND
WONDERFUL ACCOMPLISHED GO
WITHOUT REPORT

- HERODOTUS OF HALICARNASSUS

MCMXCVIII

William also gave us most of the ideas as to what to put into the capsule. I have insisted to him that in 2051, when The Hill celebrates its 200th anniversary, there will be great pressure to open the time capsule. In my opinion, at that time he will be Chairman of the Board of Trustees and it will be up to him not to allow that capsule to be opened. It should remain sealed for at least 100 years and preferably until the building comes down.

During the first year of coeducation The Hill still had an honors section in Algebra I. The Mathematics Department discontinued that section when the School went coed because we had dropped the second form and most of the Algebra I Honors section was made up of second formers. Thus we had an honors section of Algebra I only the first year The Hill was coed. I therefore always claimed to Stefany Bortz and Beckie Knight that they were the two greatest female Algebra I Honors students in the history of The Hill School. It may be true they were the only two girls, but that means they were the greatest.

I sent a letter to many Hill graduates so that it was at their colleges the day they arrived on campus. I wanted to give them last minute suggestions on being successful in college.

Rules for a College Student

These rules guarantee that you will enjoy your college experience.

1. Have a good time. Party as many nights as you can. Remember, you only go to college once.

2. Skip breakfast every morning. This allows you to stay up later at night. You can always rationalize that you do not like to eat breakfast anyway.

3. Be sure to take many class cuts. What good are they if you do not use them?

4. Since you have already studied much of the material covered in your course work, do not review. Use this extra time to enjoy yourself.

5. Do not take good class notes. You can always borrow good notes from a friend who pays attention during class.

6. Underlining or making notes in the margin of your textbooks decrease the resale value. Besides, that nonsense does not really help one remember.

7. If a course seems the least bit difficult, change courses. Always take the easy way out.

8. Never use the library. The lighting is usually bad and you do not want to be known as a grind.

9. Look out for number one only. Use anything which belongs to your roommate if you wish. Do not ask, just help yourself to all of his or her munchies. Turn your stereo as high as possible. Do not let anyone else study and possibly get ahead of you.

10. Never think about the large amount of money your parents are paying toward your college education. This will only make you feel guilty for not trying your best to earn good grades.

11. Do not let your teachers at The Hill School know how you are doing. They worked with you over the last several years but are no longer interested in you.

Follow these rules closely and it is guaranteed you will enjoy your first, and only, year in college.

A. Friend

I had many students write back that they hung the letters in their dormitory rooms as a good reminder to work hard.

Chapter X

The Last Year

There are great anxiety and anticipation at the beginning of every school year. Having completed forty years of teaching at the secondary level and this year was my last, I felt especially uneasy. Would it be different? Would I be able to keep up the same enthusiasm knowing this would be the end? Would my students react any differently? Would I become a little dejected as the year progressed? I was anticipating the year more than ever, yet I was reluctant to have it begin. There would be so much that would be new just because it was the last.

The first day of classes of my last year of teaching will forever be etched in my mind. The day was September 11, 2001. I had a free period in the morning and Mike Pentz, a fellow math teacher, stopped in the room to tell me a plane had hit the World Trade Center. I went to the student center to see the news on television. As I watched the second plane hit the World Trade Center, I could not believe what I saw. Students were gathering between classes to see the news. Our Headmaster David Dougherty handled the situation perfectly. He held the regular chapel service that morning to outline in detail what was known at that point. He emphasized that we must not let this tragedy interfere with our purpose any more than it had to. We would continue classes as scheduled. Counselors would be available all day and night for anyone, student or faculty member, who wanted to talk. I could only think back to that November day in 1973 when our dining hall and guest house, along with many administrative offices, burned to the ground. Then-Headmaster

Chuck Watson spoke to the School, saying that although this fire was catastrophic, it would not deter us from our mission. Classes would go on. The gymnasium was made into a make-shift cafeteria and school continued. The Hill is a family school and in both cases, September 11 and the fire, the family pulled together and helped one another out. We were able not only to survive but to live on and excel in all the usual areas of school life. We were quite fortunate at The Hill in that no member of a student's immediate family and no alumnus lost a life on that fateful day, 9/11/01

I started the year keeping a journal. After two weeks I stopped. It was taking too much time. I had so much to write. Things happened so fast and furiously I could not afford a couple hours of writing every night. Looking back, I wish I had forced myself to do that writing. This chapter would certainly have been a little easier to handle. However, this chapter would have grown so long that it would have become a book in itself. Maybe it could have even become a two volume book.

I was very fortunate to have some great classes for my last year. Two of my classes were non-Advanced Placement calculus classes. These sections usually consist of students who were not particularly great in mathematics but have an interest in learning math and make honest efforts. Unfortunately, one of the sections had several students unwilling to really put in an effort. The other section was reasonably good. I also had an Algebra I regular class. Although I have had many brighter students in Algebra I in previous years, this was a class with character. The students worked, and they were a happy class. I never had to pull responses out of the kids. They were always ready to take part in class discussions and were not afraid of making a mistake. Many, many times they were wrong. But doesn't one learn from his or her mistakes? I have always felt those students who take an active part in class discussions learn so much more than those who don't. Mathematics, in particular, is learned by doing. One cannot sit and watch and be able to learn facts. Unless a student really gets down and pushes the pencil, he or she will not be able to learn mathematics. I repeatedly explained to my students if they only went to the gym every afternoon, listened to the coach explain how to hit a baseball, and intently watched his demonstrations, then come game time they would strike out. It takes practice and more practice to succeed in athletics; the same is true with mathematics.

My fourth and final section was Algebra II Honors. It is probably the most difficult math course at The Hill School because it includes so much material. The syllabus for this honors course includes the traditional Algebra II topics plus the traditional pre-calculus topics. Basically, these students do two years of mathematics in one year. I was extremely fortunate to have a group that not only was as exuberant as my Algebra I class, but also included many great scholars. It may have been my best class ever. These students loved mathematics. They were so bright that they quickly comprehended the subject, and they made the classroom atmosphere glisten. This group had the ability to fool around for five minutes at the beginning of a class period and then at the words "let's begin class," become all business. Needless to say, most of these students received grades of "A."

As the end of the Fall Term approached, all I could think about was that I would never have this great sensation again. Never again would I be able to walk from home in the morning to see those leaves change color and then begin to fall. Never again would I see all those football, soccer, and field hockey games. Sure, I would be able to return in years ahead to photograph the games, but I would not really know the athletes. I would not be taking a photo of the football player who happened to be in my math class or happened to sit down to have dinner with me at night. It would be different and I knew it. I was determined, though, to get all I could out of it, one last hurrah.

The last weekend of every fall season is Lawrenceville Weekend, when all our teams compete against their Lawrenceville rivals. Each day of the Lawrenceville week a sweep broom is awarded to the captain of different varsity athletic teams. The broom passes from cross country to water polo to soccer to field hockey to football. The broom represents a clean sweep: all our teams to win. In my thirty two years at The Hill that happened only once. On Friday I was handed the sweep broom and told that the weekend would be dedicated to me. I was also, given a framed photograph that Carl Gachet had taken of me photographing a soccer game. That photo now hangs in my study at home. Although not all of our teams won that day, I will remember it as one of my most enjoyable Lawrenceville weekends.

The Winter Term is always quite dreary. It is my least favorite term in the year. First of all, the sports are indoors. I have a much more difficult time shooting indoor photography. I would much rather be outside with nice natural sunlight. For 32 years I was the official timer at all boys' basketball games, varsity, junior varsity, and thirds. Soon I would time my last game. I didn't mind keeping the score clock, but I enjoyed the game more when I was sitting in the stands. My concentration was on different aspects of the game when I was responsible for the timing, fouls, and timeouts. The fans in the stands never notice when the timer has a perfect game, but they sure do notice when he makes a mistake. I made my share of mistakes, but I think I did do a reasonable job.

Probably my biggest error occurred during a Lawrenceville game. With just a few seconds remaining, I forgot to start the clock. The Hill had already in bounded the ball and dribbled across midcourt before I realized my mistake and hit the start button. Many L'Ville fans were yelling and screaming, so the referee blew the whistle and came over to check on what had happened. I explained that I had forgotten to start the clock. That was fine. We went on from there and the mistake did not affect the score or the outcome of the game. However, after the game the opposing coach yelled and screamed as if it had been the worst thing anyone could ever do. I ignored him and simply wondered if he thought he was perfect and had never made mistakes.

My other big mistake did seem to change the outcome of a game. We were losing by one point, and I again forgot to start the clock until the ball had crossed midcourt. We went down and scored a go-ahead basket. The opposing team brought the ball down and missed a last-second shot. That coach thought I had cheated on purpose and thereafter refused to play us. He made this decision even though we had sent him a game tape so that he could see that my mistake had actually given his team a chance to win! If I had started the clock when I should have, we still would have had time to score, but his team would not have gotten off the last shot.

I thought I should do something at my last game to show it was my finale. Perhaps it was to show I was bothered that it was the end. I decided that I would not end my last game on the clock. There is an automatic horn on the clock so that it will sound at the correct instant. However, the timer must turn the horn off. That game, when the horn went off, I stood up, put on my coat, and walked out of the gym. I did not end my last game as a timer at The Hill School.

I always like the Spring Term. Probably the main reason is that it is baseball season. However, in the spring everything comes to life, especially at The Hill. I believe that the campus is the most beautiful during the Spring Term. The grounds crew does a perfect job with all the flowers and blooming bushes. The trees come to life first with their buds and then leaves. The students are able to get outside and run around, playing frisbee, shoe golf, and javelin ball. It gets lighter earlier in the morning and stays lighter later in the evening. There is just so much life on campus during the spring term.

My walk every morning to get to school followed the same path every day. Fortunately, I was able to pass by many of the flowering bushes. The sun rising in the East would just light up the flowers. Many early evenings I would return home that same way to have the sun again reflect off those same bushes from the West.

Every spring we seem to have some wild ducks land on The Dell (a small pond on The Hill's campus) and produce offspring. With so many students, dogs, and turtles around, the ducklings seem to have little chance to survive. But, the previous spring six ducklings had lived to fly away during the summer. My last spring there were two families of ducks. One had a nest by the Dell and the other had a nest about 200 yards away by the Headmaster's Garden. As a result, I was able to see these ducks waddling around the grounds on my way to work each morning. I often brought my camera just to get photos of the ducks in action. The first to give birth had eight little ducklings. When they were only eight days old we had a horrible storm. I am quite certain they went over the dam during that storm. The mother duck looked quite disheveled also, feathers askew and covered with mud. I am sure she had tried to save them. The second family had nine ducklings. They were doing surprisingly well and had lived for about six weeks, when for some reason, the mother led them away from campus. She returned a few days later, but without her young ones. I believe weed-killer put in the Dell bothered them, so she had tried to get them down to the river but had been unable to do so. We will never know for sure what happened.

The Mathematic Department held a retirement dinner for me. My colleagues rented a local restaurant on a night of the week when the restaurant was usually closed. Matt Ralston, as Chairman of the Mathematics Department, was in charge. He put together an evening I will never forget. First of all, he invited Frank deLaurentis, a good friend and former math teacher at The Hill, and his wife Betty. They both came, and I can assure you, no party is complete without Frank. Matt also invited Larry and Carol Kelly. Again, they are former Hill faculty who left to teach elsewhere. Although the Kellys were not able to come, they sent their best wishes and a few great gifts that are in my house today. One of those gifts is a night light, in the shape of a baseball, of course. So I see that light every night and am reminded of two great friends. Mr. Ralston also invited Marty and Lindi Vollmuth, two of my closest friends on campus. I was able to bring my former registrar, Kathryn Miller, to the event. Bob Parker, who had followed me as department chairman, instituted a policy of allowing single faculty members to invite a guest to the year-end dinner. Matt Ralston thankfully followed that tradition. Bob also began the policy of inviting retired faculty who live in the area to that dinner. I sure hope Matt follows that policy also. I have told him what a good idea it is. I was fortunate to have the members of the Math Department plus many more close friends attend my retirement dinner. The Mathematics Department gave me my very own Hill School baseball jersey. It was number 1 and had my name on the back. It is now displayed in the dugout in my home. That night I put it on immediately. I also wore it to the next Hill baseball game and had photos taken of me. That was a switch, me in front of the camera. My other gift from the department I am not sure I should include here. I am not sure how they pulled it off. I was given my very own license plate to put on my car. I now drive around with a license plate that reads *DLSF*.

That same week the yearbook came out and the staff was very nice to devote a page in appreciation to me for my time at The Hill. Although it had many errors of facts, I did appreciate the thoughtfulness displayed.

A few days later the School's newspaper, *The Hill News*, had an article about my time at the Hill. Evan Litvin, a young man I hardly knew, wrote an article of which I could not be prouder. Here it is.

Hill Legend Retires

After 32 years of faithful service as a mathematics instructor, Mr. Willis J. Pierre will retire at the end of this year. "It's been fun, I really have enjoyed it…the people I have met, students and faculty. It really has been great," says Mr. Pierre. He has been a friend and mentor to nearly all that know him. He is there with a story to tell and an ear to listen. Mr. Pierre is known to nearly all as one of Hill's most enthusiastic and famous mathematics instructors. "Willis has been one of the most 'kid-oriented' teachers I have ever had the privilege of working with," says fellow math teacher Mr. Pentz. He is also known as the "guy behind the camera" at most sports events on campus and for his vast sports memorabilia collection. Mr. Pierre has been a great part of the community during his time here and will be greatly missed after he leaves.

Willis Pierre, a native of Princeton, New Jersey, taught for nine years at Lake Forest Academy in Illinois before coming to Hill. He liked the school, but wanted to be closer to his hometown. So, he used the trusty *Peterson's Guide to Prep Schools* book and wrote to fourteen boarding schools in the North-east. It was by chance that he met with then headmaster, Archibald Montgomery. Mr. Pierre was told to call Mr. Montgomery the next time he was on the East Coast. So, during his midwinter vacation in Princeton, he attempted to call Hill but he kept getting the 'this line is no longer available' message. Mr. Montgomery eventually called him the day

125

before he planned to go back to Illinois, and by chance, on the way to the Philadelphia Airport, Mr. Pierre finally made it to the campus of The Hill School. He loved it. By the time Mr. Pierre left campus, he knew this is where he was meant to be.

Since then Mr. Pierre has developed a reputation for kindness and enthusiasm towards all parts of life ranging from the classroom to the sports field. He truly enjoys waking up every day and going to teach math class. Since he was a kid, Mr. Pierre has loved math. It was one of his childhood math teachers that influenced him to become a teacher himself. Mr. Pierre has had such classroom experiences as teaching a Geometry Honors course that developed its own theorems and ideas. This book went on to become a marvel to geometry teachers across the country. He was also picked as a teacher to be focused on for a Disney Channel special that aired on cable television.

Mr. Pierre has always been a fan of all sports, although his true passion lies in baseball. Around the same time he began liking baseball, he began collecting sports memorabilia. Today his sports memorabilia collection is amazing. His collection includes autographed baseballs, seats from the old New York Giants stadium, and a Hill baseball coaching uniform worn by Ty Cobb. Recalling sports memories during his time at Hill is a pastime Mr. Pierre truly enjoys. Tales of game-winning homeruns, last-second three-point shots, and game-winning field goals fill Mr. Pierre's memory.

Throughout the past six years, Mr. Pierre has also become a well-known photographer around campus. He is at nearly every sporting event, taking pictures. He started out with this hobby by taking the Hill's photography course under Mr. Carl Gachet. Mr. Pierre attributes all of his knowledge on this topic to that course and Mr. Gachet. If you get a chance, the next time you are on the third floor of the Academic Center, take a minute to admire those amazing photographs on display covering the entire hall. After Mr. Pierre retires, all of his photographic work will be constantly updated, but remain here as his legacy to future Hill students.

"The only thing I am not going to miss about Hill is the meetings!" says Mr. Pierre with a smile. "I am really going to miss just teaching, and giving extra help." Mr. Pierre's retirement plans are still undecided except for maybe a trip to see his favorite baseball team, the Giants, play in spring training and for a few more trips to the Baseball Hall of Fame each year. Mr. Pierre has bought a house in Trappe, PA, which is relatively close to Hill where as of now he plans to spend his post-Hill years. He is also considering teaching part time, "but only if there are no meetings!" he jokes. Over the next few years while he still knows most of the students, Mr. Pierre plans to spend time taking pictures at sports and watching the games.

Mr. Pierre would just like to add, "I would just like everyone to know how much I have enjoyed teaching at The Hill School... it's been fun.... I think the best way to explain it is that Hill is my field of dreams."

On my last day of teaching classes I was quite apprehensive. Would anyone realize this was the end of my teaching at The Hill? It was a beautiful morning. I made my usual walk to school and arrived by 6:30 a.m., so I could sit in the dining hall with a cup of coffee and read my newspaper. Slowly students and faculty arrived as usual and we had our breakfasts. No mention was made that this was my last day.

127

At The Hill School, seniors (sixth formers) do not have to take final examinations. Exams are given following graduation weekend. Underformers have classes on that Monday and exams during the remainder of the week. Since two of my classes were calculus with only sixth formers, I had only two classes on my last day. I had planned to do a great amount of work during my free periods. Little did I know that I would not have a free period. My day would be full and quite eventful.

My first class that morning was my Algebra I class. All went well as I reviewed everything that would be on the final examination. Then I gave my students the test they had taken the first day of class. Of course not a single person had remembered seeing the problems before. With about five minutes left in the class, Michelle Khwarg, one of the more outgoing members of the class, stood up and walked to the front of the room. She had a very nicely wrapped package, which she presented to me. Michelle thanked me for a great year and said that the class hoped I would have a good retirement. It so happened that a few weeks earlier the class had asked if they could take a picture of everyone including me. Kendall Taylor had brought a camera, put it up on a desk, aimed it, set its timer, and ran to get into the group photo. Little did I know that the real reason they wanted the photo was to present it to me on the last day of class. They had painted the frame with math symbols and stars and had written "from your favorite Algebra I class" on the frame. Even the card was a special gift in itself. The outside of the card read "Retirement is a time when one can look back over one's career with satisfaction and say 'I've done well!' It is also a time when one can look to the future with anticipation and say…" The inside of the card read "To hell with Mondays, I'm sleepin' in!" Each member of the class had written a short note on the card. Here are a few comments these third formers wrote.

Catherine Eisenmann wrote, "It's been great! Thanks for everything. Your class was always an enjoyment! Congratulations."

"Thanks for the great year. It was a privilege to have you as a teacher. Have a great retirement," signed by Evan Malone.

Caren Beck, whose brother I had had the year before, wrote, "Thank you for a great year. It was awesome to come and look at all the pictures and learn a little math on the side. I'm glad you were my teacher this year. Keep in touch and come back to visit when you have time." I should add that Caren did learn a little math: she won the prize as the top Algebra I student in the School.

Paul Fusco, one of the hardest working students in the class, wrote, "Thanks for everything this year, especially all those nights of extra help. Hope to see you around next year taking pictures of my basketball and lacrosse games. Time to relax now!"

Quiet Katie Robinson penned, "Thank you for a great year of math! I learned so much and had fun doing it. I'll miss you next year!"

Finally, Michelle Khwarg wrote at the end of the card, "First, thank you very much for everything (jokes, enthusiasm, teaching and more). I personally think you can say 'I have done well.' I know you never sleep in. I do! Haha. I love your photos." It is true that Michelle missed class one time because she had overslept.

When the class ended, each student came forward to personally thank me before he or she left the room. I sat down to do some work. In less than five minutes the first of many students walked in to hand me a gift and thank me for the year. It was almost as if they had this procession timed so that I was not to have more than five minutes alone in the room. By the end of the day, I had quite a collection of gifts to cart home. Among them were many beautifully inscribed books, and to add to my bobbin head doll collection were a Barry Bonds, a Cal Ripken Jr., a Tony Gwynn, a Sammy Sosa, and an Ichiro. One young man stopped in to say that he had an autographed baseball for my collection, but he had left it home. A couple days later he brought the ball in and handed it to me. It was certainly one I did not have in my collection. Very clearly written across the sweet spot was the name "Jamie Requa." That ball is on display in my computer room now along with quite a few of my other autographed baseballs. However, it is probably the only baseball I own autographed by one of my students.

My other class scheduled for that day was to meet immediately after lunch. It was my Algebra II Honors class. I thought that was appropriate, that perhaps my best class ever was to be my last. I walked up to my classroom about ten minutes before class time and saw quite a crowd of students around my door. As I approached, they claimed that they knew this was my last class and they wanted to be a part of it. Could they attend? Obviously, I did not have enough chairs for them, but they were willing to sit on the floor or stand in the back of the room for this class period. Several others entered the room and stood in the back after class had begun. I don't even remember what I taught, if anything. I was awestruck by the fact that students not in the class wanted to attend. I assume I went over materials that would be on the examination. With only a few minutes left in the class, I thought I would play a game so that I could end my teaching career to a standing ovation. So I tricked the class with a little test that ended with all of them standing up and clapping. I thanked them and claimed that class was over. One by one everyone in the room, students and visitors, came forward to thank me and say good-bye. It was quite a moving experience for me and one that I appreciated for my last class. I went home, sat down and relaxed.

Over Alumni Weekend in June, The Hill has a special program to honor those faculty members who have just completed their twenty-fifth year at The Hill and those who are retiring. The person being honored can choose anyone to give a short speech about him or her. There was no question in my mind whom I wanted. I really feared even asking because I thought there was no way he would come all the way to The Hill School from Virginia just to talk about me. I finally decided that the worst thing that could happen was that he would say no. I scraped up enough courage to ask and was flabbergasted with his reply. "That would be an honor; of course I will do it." So I was having Chuck Watson, my former headmaster, return to our school to speak on my behalf. I cannot begin to list all the great things he did for me while he was the Headmaster of The Hill School. I can say that I doubt very much that I would have remained so many years at The Hill, if I had had any other headmaster. Here is what he said at that ceremony.

I should not be here today. But then if something I did had not been ignored, Willis would probably not be here today either. He is here, however, and being the generous and forgiving man that he is, he invited me to join him.

You see this all goes back to that winter day in 1970 when Willis came to The Hill to be interviewed for a position in the Mathematics Department. As Director of Studies at that time, I was scheduled to meet with Willis. To my forever embarrassment, it seems that, taking leave of the few brains I have, I wrote on the interview assessment form that I just was not certain and that perhaps we should keep on looking. Fortunately, wiser heads, for which I have always been thankful, chose to ignore my comment. How totally I was able to misjudge this man, wanting to return to the East after a distinguished career in Illinois at Lake Forest Academy, who embodied that mixture of intellect, enthusiasm, loyalty, knowledge, and common sense that every boarding school longs to have, I have never understood. I apologize.

I realize now, however, that for the next twenty some years, Willis extracted his pound of flesh or maybe I should say his pounds of quarters. Each May at the Strawberry Festival Willis runs a game that, to speak bluntly, closely resembles the old shell game played on city streets--although, it is the Pierre version of it. Since all Festival proceeds go to charities, whatever name Willis had for his operation carried official administrative approval. With a gleam in his eyes as he saw the biggest dupe on campus approaching, Willis would watch me plunk down my first quarter. Then he would go to work. Students to my left would win; students to my right would win; parents, faculty, staff roaming by to play would win. Would I? Only when it appeared out of frustration I might quit playing! After running through five dollars of quarters and having no more, the always graciously encouraging Willis would tactfully suggest that I might like to drift over to the cashier's table and get five more dollars in quarters, which I would do. After losing them also and giving up, Willis would cheerfully say, "Gee, too bad. Better luck next year."

I understand that Willis will be living in Trappe, a small town close by. What a perfect name for Willis' game playing. I wonder if I should warn the Trappe fire halls and churches never to let him into their bingo nights.

During his thirty-two years with The Hill, Willis has held an array of critical positions, discharging the responsibilities of each with fairness, skill, and gentle humor. Little wonder why he has long been regarded as an extraordinarily effective school administrator. He is also recognized nationally for having assembled one of the finest baseball card collections in private hands.

Above everything, however, is his love for and enjoyment in teaching mathematics. Nor is it just The Hill School and countless numbers of his students who recognize the excellence of his classroom teaching; the world outside does also, as evidenced by the grants Willis has received from the National Science Foundation, the Endowment for the Humanities, and his articles published by the Endowment and the University of Delaware.

Of them all, however, and in my opinion at least, the one that surpasses the others was the grant Willis received about eighteen years ago from the National Mathematics Trust to have his Geometry Honor students write their own geometry text. The finished book, entitled *Geometry According To Us,* was worthy of national publication, and probably would have been were it not that in most public schools geometry is today given cursory treatment, if even that. In other words, the market for the remarkable project was not broad enough.

Let us therefore take a few minutes to slip inside this master teacher's classroom and listen to him lead a geometry class through an intricate proof. With paper, pencils, erasers, compasses, rulers, protractors and aspirin students are sitting at the ready as Willis speaks.

See the famous Pierre theorem
And how it's applied.
It states that on weekends
Parallel lines will collide.
There are those who claim that

My theorems all rot.
But I know that I'm right
And I know that they're not.
Hook the straight line marked A
To the curved line marked purple,
Now your first step's complete,
You've constructed a circle.
Next keep your eyes searching madly
In wide open focus
Or you'll fail to discover that thing
That's called locus.
Oh, please don't complain that
My teaching's obtuse,
Or you'll never track down your hypotenuse.
And if you say the geometry's
Not lucid,
Well, don't blame me, just
Blame old Euclid.
So take pride in this proof
Of mathematical flair,
In geometry according
To Willis Pierre.

I was not aware until Willis told me, followed a few days later by the announcement from the School, that others are also being honored this morning. Each of you has been a friend to Biddy and me for more than two and three decades, and we sincerely thank you for that friendship.

May I therefore say to you Jim Long, Jim Finn, Wayne Curtis, Pete Maynard, Wayne Bell, and Willis Pierre that I hope from this day forward you shall always sail ventis secundis--with prospering winds.

As I walked off campus that evening I could reflect back over forty-one years of great fun and pride. However, this one day made a difference. Not only did Chuck Watson return for me, but my cousin, Anne Gallagher, my closest living relative, her husband and son also came. I had so many alums come up to congratulate me and to thank me for all I had done. So I walked home thinking that yes it had been forty-one years well spent and definitely worth all I had put into them. After arriving home and finally being able to relax for the first time all day, I read the program that The Hill School had printed for that occasion. Here is what it had to say about me.

"Willis Pierre joined The Hill's Mathematics Department in the fall of 1970 after teaching mathematics at Lake Forest Academy for nine years. He earned his B.A. from Washington and Jefferson College and his M.A.T. in 1968 from Washington State University.

For 32 years Hill students have benefited from Willis' innovative ways of teaching mathematics. With his guidance, Willis' students published a book entitled *Geometry According to Us*. Mr. Pierre has written problems for "Resource Problems to Enhance the Teaching of Mathematics" published by the University of Delaware; an article on "Morley's Theorem," several short pieces and letters for "The Mathematics Teacher"; and articles and letters for "Sports Collector's Digest." In addition to teaching, Willis served as a dorm parent, coach, and dedicated fan for his student athletes, attending athletics events, usually with his camera in hand. He has served as timer for varsity basketball games for all of his 32 years at The Hill and over 100 of his photographs are framed and hanging in offices and hallways of the School.

From 1976 to 1986, he served as Hill's Director of Studies before taking over the chairmanship of the Math

Department that same year. Willis has been a supervisor of the dining room, in charge of dining room seating, and director of the workjob program for 16 years. He also has served on various campus committees including Executive Committee, Scholarship Committee, Academic Council, and Faculty Council.

He has received awards for his outstanding teaching, including the Kohl Teaching Award, Tandy Corporation Teaching Award, and The Hill School's Master Teacher Award. In 1995, Willis was one of 60 teachers across the United States selected as subjects for a Disney Co. film entitled *Salute To The American Teacher.* Mr. Pierre has been named to four editions of *Who's Who Among America's Teachers* and has been chosen this year for the fifth time."

After a short break in the late afternoon it was time to go back on campus for the alumni dinner. I believe that night at least one hundred alums came over to say a few words. I was certainly quite moved by all the attention. I walked home that evening exhausted but exhilarated. I will remember that day for the rest of my life.

Epilogue

It is impossible for someone who has enjoyed teaching so much to pack his belongings one day and just walk away. Once teaching is in your blood and you really enjoy the experience, you do not want to stop. I am tired of all the busy work that has to be done, but I am not tired of teaching. I hope to be involved with teaching to some extent during my retirement years.

I am extremely fortunate to have had so many years at The Hill School. The School was extremely good to me for my entire thirty-two years. I never dreamed that it would continue to be so kind after I retired. I have been allowed to maintain an office on campus. Since many students and parents are interested in purchasing my photographs, the office gives me an ideal meeting place. I keep most of my photography work in that office, where I review and file the photos and decide which ones to enlarge. I have become more aware, as space gets smaller, that I must discard many photos. A number of years ago Carl Gachet told me that if you get one or two great photos out of a roll, you are doing well. I have not gotten to the point where I can discard all the non-great photos, but at least now I can justify throwing away the bottom twenty percent. Students know I am in the office two or three days a week and stop in for a chat or, on occasion, extra help.

It is also quite fortunate that I was able to acquire a part-time teaching position at Montgomery County Community College. I was a bit leery at first because I did not know exactly what to expect. So far, teaching there has been a tremendous experience. In the Fall Term I taught two sections: Intermediate Algebra and

Math Applications. The Intermediate Algebra course covered material I had taught at The Hill School, mostly in Algebra II. The Math Applications course had very little material I had taught before. The beginning of Math Applications was a good review of algebra which, of course, I had taught previously. The rest of the course involved banking and probability. I have always enjoyed probability, but it was not part of the curriculum at The Hill School. As I mentioned before, I love problems with unexpected outcomes. I introduced these kinds of problems into this class. They made for a much more interesting course.

One of my favorite problems involves a poker hand. Suppose you are dealt the spade king, queen, jack, and 4 along with the diamond 10. Your most likely choices for discarding are the diamond 10 to try for a flush or the spade 4 to try for a straight. In poker the flush has a higher rank than the straight. Therefore, you would expect the straight to be the easier hand to catch. In this situation you would be wrong. Not only is the flush a better hand, but you are more likely to catch the flush than you are the straight.

My other favorite probability problem is the birthday problem. If I stop the first 100 people I see in greater downtown Pottstown and ask them for their birthdays, months and days only, what is the probability that I would have a match? After going through the problem and seeing that I almost certainly would have a match, I ask the class to find the number I would have to stop to make the probability one-half. The solution is just 24 people. That so few people are needed amazed my students. The reader may like to check a group of friends and see what the results are in that specific case.

The banking problems caused some interesting discussions. When studying mortgages, my students were shocked by how much a person ends up paying for a house over a thirty year period. They also were dismayed to learn how little of the first month's payment goes toward the house since most of it is only interest payment.

I found students at The Hill and students at the Community College to be quite different. I must admit I was much more impressed with the students at the Community College than I had thought I would be. They were goal-oriented and worked hard. Most of them held down at least part-time jobs while studying. Seldom did they fail to turn in assignments, and when they did, they brought the late work the next time we met. At The Hill, extra help was a tremendous resource for the students. I continued this practice at the college and was astounded by how much the students wanted to take advantage of the offer. It was gratifying to see that almost every student really wanted to learn. I do not mean that students at The Hill do not work and do not learn. However, at The Hill I did have many students I had to drag and cajole to do the work. Many of them would not turn in assignments if they could find good excuses not to do them. I spent an extraordinary amount of time having to coax students to work. Of course, many students at The Hill not only were very bright but also had a great interest in learning.

Each of my classes met for one and one-half hours twice a week. So I taught for three hours every Tuesday and Thursday mornings. It was a change of pace for my retirement and I could still enjoy teaching. I tried to work Monday, Wednesday, and Friday mornings on this book. However, when I got up in the morning and did not feel like writing, I didn't. Retirement is great! You can do whatever you want.

There are a great many retired employees of The Hill School who still live in close proximity to Pottstown. I wanted to capitalize on this situation and form an organization of those who have an abiding interest in The Hill. I asked a few of the other former faculty members to help me in this project. From the very beginning Wayne and Marsha Bell, David and Ginger

Giammattei, and Tom Ruth were very good at giving up some of their time to help make this project work. Later Kathryn Miller and Jim Taylor joined our early morning breakfast meetings. We plan to take this first year to organize the group and decide exactly what we want for our future. I hope that next year we will be a very active group. I believe it is not only worthwhile for us, the former employees, but also, in the long run, beneficial to The Hill School. In June, exactly one year after my retirement, the first meeting of the *Hillbackers* will occur. There has been a great response from people who are interested in joining our organization. Hopefully, *Hillbackers* can support The Hill and make it a stronger institution and at the same time permit us to continue to be a part of the School.

I am also extremely lucky to have found a nice townhouse which I adore. It is only ten miles from the Community College and The Hill School. I feel fortunate because I looked at so many houses before I finally bought the perfect home for me. The home is large enough to display all my collections and also is surrounded by a large, grassy open area. Having worked in a boarding school all my life, I have never had to mow lawns and shovel snow. I was not about to start doing these chores in retirement! I have been extremely blessed with great neighbors and the nice atmosphere of a college town setting.

Teaching is quite a rewarding profession. I am glad that I chose it. There are probably some things I would do differently, if I had to do them over again, but not many. I am sure I have been a failure to a few students, but by the visits and letters I receive daily, I know that I have had a positive impact on many students' lives. Through e-mail I continue to bug the DLSFs. *Saturday Night Survey* is going stronger than ever. I continue to be busy, very busy. I often wonder how I had time to keep up with a full-time teaching position. There is never a dull moment.

If the reader is a young person trying to decide on a career, consider teaching. You will not only teach, but you will learn daily for the rest of your life.

Finally Brethren
Whatsoever things are true,
Whatsoever things are honorable,
Whatsoever things are just,
Whatsoever things are pure,
Whatsoever things are lovely,
Whatsoever things are of good report;
If there be any virtue,
And if there be any praise,
Think on these things.

Philippians 4:8

Appendices

Appendix A

Answers

Page 34 -
 The score before the game begins is 0-0.
 The desk can't jump.

Page 43 -
 The number of letters in each word equals the digit value
 for pi.

Page 45 -
 $8 + 8 \div 4 \cdot 2 = 12$
 "is a brother of" is not symmetric but is transitive.

Page 46 -
 Let the distance be x.
 The time going is x/60 and time returning is x/40.
 Average rate is always total distance divided by total time

$$\frac{x + x}{\dfrac{x}{40} + \dfrac{x}{60}} = 48 \text{ (B)}$$

Page 46 -
 Assume that they both have 160 hits in 400 at bats.
 Frank First's final average is 165/406 = .4064
 Steve Second's final average is 168/412 = .4078

Page 47 -
$$(1)(.90) + (x)(0) = (1+x)(.50) \quad x = 4/5$$

Page 47 -
$$\frac{x}{30} + \frac{x}{360} = 1$$

Solve for x. x = 360/13 or approximately 27.7 minutes.

Page 47-48 -
   Boat

Page 49 -
   Noel – No L

Page 50-51 -
1. sine
2. tangent
3. cosine
4. secant
5. IV (ivy)
6. top half of XII
7. wrong

Page 52 -
   Bend the paper to form a doughnut shape. Then draw the lines. In mathematics this shape is called a torus.

144

Pages 54 -
    $(ice)^3$ = ice cube
    log(cabin)
    $(ice)^3$
    ln(cabin) = natural log cabin
    ln(cabin) + C = houseboat

Page 101 -
    F(1) = 1, f(2) = 1, f(3) = 2, f(4) = 3, f(5) = 5 etc.

Appendix B

Photographs

Front Cover

Patrick Lundquist found studying in the Ryan Library. Have you ever seen someone so happy while studying? This is not a posed shot. "Studying With A Smile" has been published by The International Library of Photography in *Best Photos of 2003*.

Jonathan Clarke is photographed in a tree next to the Alumni Chapel. Jonathan was often used for posed shots like this one. This is an award winning photo.

Ainsley Becnel makes a successful pole vault at The Hill School track. This photo has won two awards.

The Hill School dining room has the tables set for a formal sit-down meal.

Back Cover

Willis J. Pierre wears the baseball shirt given to him by the Mathematics Department at his retirement dinner. He set up the camera and then ran to get into the photograph.

The license plate presented to Willis J. Pierre at his retirement dinner. The little girl is Nicole "Niki" LoBiondo, Willis' next door neighbor.

Appendix C

Acknowledgements

Finally, I must thank a number of people who helped so much in making this book a reality.

To Kathryn Miller, Jim Long, and Michael Hoch, whose proofreading made this book readable for you, I give my heartfelt thanks. I hate to admit the number of errors they found in the original pages. I know they spent much more time making corrections than any proofreader would need to do for a professional writer. I thank all three individuals for their careful reading and suggestions. A special thanks must go to Mike Hoch, my only non-Hill School proofreader. He was able to see things the rest of us took for granted. I know someday he will be a great teacher. I hope he chooses mathematics as his subject.

I thank Rosie and Peter Shiras not only for their many corrections and suggestions, but especially for their great encouragement to complete the project.

Larry Kelly volunteered to write about me for this book. I thought I was being very coy with my requests for permission to use answers to questions on *SNS*s. Only Larry figured out immediately that I must be writing a book and he quickly volunteered to help. I guess he knows me better than I thought. It probably is a result of all those holidays I spent in his home. Larry and Carol always treated me like one of the family. I should have had him proofread all the mathematics in the text. Thanks.

George Kovac I thank for his many kind words for this book and over the years. He was always ready to attend a sports event with me whenever I asked. George enjoyed teaching more than anyone I know. I regret that he did not have many more years with us at The Hill.

Mike Pentz was my trusted confidant my entire career at The Hill. He always knew the right words to say in any situation. Because I taught in a room next to his, I could conveniently wander over to get all my questions answered. Thanks.

Morris Publishing I must thank for pointing out all the little details that need to be done before a book goes to print. A first time author, I would never have thought to do many of those things. I also thank Morris Publishing for helping me to secure a copyright, an ISBN, a bar code and a Library of Congress number.

You, the reader, I thank for having the interest to pick up this book and spend a little time with it in your hands. I hope you enjoyed the experience of reading *DLSF*.

Index

FOIL, 50-51
Forbes, Malcolm S., 30
Ford, Betty, 24
Ford, Gerald R., 24, 30
Forgettables, The, 59
Fort, Pat, 112
Frank, Clinton E., 29
Friedman, Milton, 31
Fullmer, Gene, 31
Fulmer, Paul, 88
Fusco, Paul, 129

G

Gable, Dan, 31
Gachet, C.C.F., 31, 121
   126-127, 137
Gallagher, Anne, 85, 135
Gallagher, Philip G., 31, 71,
   72, 74, 76, 81, 85-86, 135
Gallo, Ernest, 30, 32
Gallo, Julio, 30, 32
Garagiola, Joe, 24, 29
Garner, James, 25
Gasparro, Frank, 31
Gates, William H., 30
Gauthier, George, 5-6
Gavin, Pat, 81
Geometry According To Us,
   57-59, 133, 135
Gerasimowicz, Nick, 84
Giammattei, David, 139-140
Giammattei, Ginger, 139-140
Giangiulio, Derek, 105
Gibbs, Joe, 29

Giel, Paul, 29
Gizzi, Tom, 85
Glanville, Jerry, 29
Glenn, John, 30
Gola, Thomas J., 30
Goldwater, Barry, 24
Goode, W. Wilson, 30
Goodrich, Gail, 30
Goodyear Blimp, 19
Gore, Al, 33, 87
Graham, Otto, 29
Gray, Pete, 33
Greenlee, Cam, 83, 85
Grier, Rosey, 29, 33
Griffin, Archie, 29, 32
Griffin, Merv, 25
Groon, Daryl, 89
Groon, JC, 76, 81
Groten, Nils, 85
Groza, Louis R., 29
Gwynn, Tony, 130

H

Haddix, Harvey, 29, 32
Hagge, Chris, 75
Haig, Alexander M., Jr., 30
Hall, Monty, 25
Handwerk, Taylor, 84, 105
Harmon, Tom, 29
Harris, Sara B., 57
Harvey, Paul, 25
Havlicek, John, 30
Hayakawa, S.I., 30
Helms, Jesse, 30

154